Applied Analysis of Ordinary Differential Equations

Applied Analysis of Ordinary Differential Equations

Special Issue Editor

Sanjeeva Balasuriya

MDPI • Basel • Beijing • Wuhan • Barcelona • Belgrade

MDPI

Special Issue Editor
Sanjeeva Balasuriya
University of Adelaide
Australia

Editorial Office
MDPI
St. Alban-Anlage 66
4052 Basel, Switzerland

This is a reprint of articles from the Special Issue published online in the open access journal *Mathematics* (ISSN 2227-7390) in 2018 (available at: https://www.mdpi.com/journal/mathematics/special_issues/Applied_Analysis_of_Ordinary_Differential_Equations).

For citation purposes, cite each article independently as indicated on the article page online and as indicated below:

LastName, A.A.; LastName, B.B.; LastName, C.C. Article Title. *Journal Name* **Year**, *Article Number*, Page Range.

ISBN 978-3-03921-726-7 (Pbk)
ISBN 978-3-03921-727-4 (PDF)

Contents

About the Special Issue Editor

Sanjeeva Balasuriya was awarded his PhD in applied mathematics from Brown University, and has since held academic position in three continents. Currently at the University of Adelaide (Australia), his research comprises three aspects of differential equations—theoretical, numerical, and modeling—often in combination. He has won numerous awards for his research, e.g., Future Fellow of the Australian Research Council, Endeavour Research Leadership Award from the Australian Department of Education and Training, J.H. Michell Medal from the Australia/New Zealand Industrial and Applied Mathematics Society.

Preface to "Applied Analysis of Ordinary Differential Equations"

This book is a compilation of articles from a Special Issue of Mathematics devoted to the topic of applied analysis of ordinary differential equations. My original goal in editing the Special Issue—and now the book based on this—is in traversing the interesting boundary between the well-established theory of ordinary differential equations and the applications of these. There are a multitude of applications in the literature which are based on numerical work because differential equations are often intractable to analysis. These are useful and have been widely used, but often possess many parameters which are usually estimated in some way. Differential equations are, however, notorious for their sensitivity to parameters and, therefore, results from these approaches are not necessarily robust or absolutely compelling from the viewpoint of applications. On the other hand, differential equations for which analysis is possible across all parameter values are, by their very nature, simple, and are therefore less realistic. What is of interest to me is the intermediate ground, in which the differential equations contain sufficient physics/biology to be relevant and, yet, are able to be analyzed in certain ways if clever approaches are followed, leading to more confident results.

The chapters of this book comprise several articles within this realm. The applications include molecular physics, epidemic modeling, cardiac arrhythmia, and nonlinear thermostats. The authors use physical, biological, or other applied insights in obtaining mathematical models via differential equations. Analysis is then used to relate mathematical properties to the applications at hand. I hope that you enjoy these chapters.

Sanjeeva Balasuriya
Special Issue Editor

mathematics

MDPI

Article

A Surface of Section for Hydrogen in Crossed Electric and Magnetic Fields

Korana Burke [1,*,†] and **Kevin Mitchell** [2,†]

1 Department of Mathematics, University of California, Davis, CA 95616, USA
2 Department of Physics, University of California, Merced, CA 95344, USA; kmitchell@ucmerced.edu
* Correspondence: kburke@ucdavis.edu
† These authors contributed equally to this work.

Received: 12 August 2018; Accepted: 19 September 2018; Published: 29 September 2018

check for
updates

Abstract: A well defined global surface of section (SOS) is a necessary first step in many studies of various dynamical systems. Starting with a surface of section, one is able to more easily find periodic orbits as well as other geometric structures that govern the nonlinear dynamics of the system in question. In some cases, a global surface of section is relatively easily defined, but in other cases the definition is not trivial, and may not even exist. This is the case for the electron dynamics of a hydrogen atom in crossed electric and magnetic fields. In this paper, we demonstrate how one can define a surface of section and associated return map that may fail to be globally well defined, but for which the dynamics is well defined and continuous over a region that is sufficiently large to include the heteroclinic tangle and thus offers a sound geometric approach to studying the nonlinear dynamics.

Keywords: surface of section; transport; heteroclinic tangle

1. Introduction

For many dynamical systems, geometric structures lying within their phase spaces provide deep insights into their behaviors [1–3]. In particular, one class of such structures consists of homoclinic and heteroclinic tangles, which have played a crucial role in studying chaotic transport and mixing [4–11]. For dynamical systems defined by ordinary differential equations (ODEs), it is typically the case that such tangles are easiest to study when there exists a "good" surface of section (SOS) which allows one to define a continuous Poincaré return map. For two-degree-of-freedom Hamiltonian systems, a SOS is a two-dimensional surface in phase space and a heteroclinic/homoclinic tangle consists of one-dimensional stable and unstable manifolds within this surface. In many cases it is challenging, or even impossible, to define a good SOS that captures all of the dynamics of the system in question. In this paper, we will consider one such system: the dynamics of a hydrogenic electron in externally applied perpendicular (crossed) electric and magnetic fields. Though there appears to be no truly global SOS, we can define a SOS and associated Poinaré map over an area that is large enough to encompass a heteroclinic tangle that controls the ionization process.

Chaotic ionization of a hydrogenic atom in crossed electric and magnetic fields has been of scientific interest for many years [12–18]. Previous work on this problem focused on studying periodic orbits and developing closed-orbit theory in order to explain the photo-absorption spectra [19–21]. Periodic orbits were also used to construct the action variables and obtain a semiclassical torus quantization [13]. More recently, the crossed fields problem has been examined from the perspective of classical monodromy [12,16]. The electron's classical motion resembles the motion of the Moon in the Sun-Earth-Moon three body system [22], and so this system has also been considered a stepping stone to understanding escape in the classical gravitational three-body problem.

For the case in which the electric and magnetic fields are parallel, it is relatively easy to define a global SOS upon which the Poincaré map is well defined and continuous everywhere. For the crossed fields case this simple construction fails to produce a good global SOS and, to the best of our knowledge, a good global surface of section does not appear in the literature. Indeed this presents one of the major challenges to studying chaotic ionization in this case. In this paper, when we say a "good" SOS we mean an SOS that intersects all trajectories (excepting a set of measure zero) and on which the Poincaré return map is well defined and continuous everywhere, i.e., the Poincaré map is a homeomorphism of the SOS. For such good SOSs system trajectories never intersect the SOS at a tangency as this would lead to a discontinuity in the map. We will present a prescription for defining a SOS that is not truly global or continuous but is nevertheless "good enough" in that it captures all the major hallmarks of chaotic dynamics including turnstiles that govern the ionization process.

This paper is organized as follows: In Section 2 we describe equations of motion for a hydrogenic electron in crossed fields; In Section 3 we present a prescription for finding an SOS; Section 4 concludes by finding a periodic orbit and its corresponding tangle and turnstile, which are responsible for the ionization process.

2. Electron Equations of Motion

Consider an electron confined to a two-dimensional plane with a uniform magnetic field **B** oriented perpendicular to the plane. The resulting electron trajectory will be a circle. Adding an electric field **E** oriented perpendicular to the magnetic field changes the shape of the trajectory from a circle to a cycloid, i.e., the motion is a combination of the circular trajectory and an $\mathbf{E} \times \mathbf{B}$ drift. Further inclusion of a $1/r$ Coulomb potential changes the electron dynamics from regular to chaotic. This is the scenario considered in the current paper. Orienting the magnetic field in the $\hat{\mathbf{z}}$ direction and the electric field in the $\hat{\mathbf{x}}$ direction, the electron Hamiltonian in atomic units ($e = m_e = \hbar = 1$) is [14]

$$H = \frac{p^2}{2} + \frac{B}{2}L_z + \frac{B^2}{8}\left(x^2 + y^2\right) - \frac{1}{r} - Fx, \tag{1}$$

where $F = |\mathbf{E}|$ is the electric field strength, B is the magnetic field strength, and L_z is the z component of the electron's angular momentum. Equation (1) assumes a fixed infinitely massive nucleus. Since the magnetic field is oriented in the $\hat{\mathbf{z}}$ direction, it is coupled to L_z by the $BL_z/2$ term in the Hamiltonian. Since we restrict the electron motion to the xy plane, $p^2 = p_x^2 + p_y^2$, and $r = \sqrt{x^2 + y^2}$. Finally, since the Hamiltonian is independent of time, the total energy is conserved in this system.

To regularize the Coulomb singularity at the origin, we introduce the parabolic coordinates [13,21]

$$\begin{aligned} u &= \pm\sqrt{r + x}, & v &= \pm\sqrt{r - x}, \\ p_u &= vp_y + up_x, & p_v &= up_y - vp_x, \end{aligned} \tag{2}$$

where u and v are the new position variables and p_u and p_v are their corresponding conjugate momenta. We take the range of u and v to be $(-\infty, +\infty)$, which represents a double cover of the physical xy configuration space. We now define a transformed Hamiltonian h, as

$$h = 2r(H - E), \tag{3}$$

where E is the electron energy, so $h = 0$ for physically relevant trajectories. Combining Equations (1)–(3), the transformed Hamiltonian h is

$$h = \frac{1}{2}\left(p_u^2 + p_v^2\right) + \frac{B}{4}\left(u^2 + v^2\right)(up_v - vp_u) + \frac{B^2}{32}\left(u^2 + v^2\right)^3 + \frac{F}{2}\left(u^4 - v^4\right) - E\left(u^2 + v^2\right) - 2. \tag{4}$$

The transformation of the Hamiltonian in Equation (3) also transforms the time variable. The relationship between the physical time t and the transformed time s is given by

$$\frac{dt}{ds} = u^2 + v^2. \tag{5}$$

In this new coordinate system, the equations of motion are

$$
\begin{aligned}
\dot{u} &= p_u - \frac{B}{4} v \left(u^2 + v^2 \right), \\
\dot{v} &= p_v + \frac{B}{4} u \left(u^2 + v^2 \right), \\
\dot{p}_u &= \frac{B}{4} \left[2uvp_u - p_v \left(3u^2 + v^2 \right) \right] - \frac{3B^2}{16} u \left(u^2 + v^2 \right)^2 + 2Fu^3 + 2uE, \\
\dot{p}_v &= \frac{B}{4} \left[-2uvp_v + p_u \left(u^2 + 3v^2 \right) \right] - \frac{3B^2}{16} v \left(u^2 + v^2 \right)^2 - 2Fv^3 + 2vE,
\end{aligned}
\tag{6}
$$

where the overdot represents differentiation with respect to s.

Finally, we introduce the vectors $\mathbf{p}$, $\mathbf{w}$, and $\mathbf{A}$

$$\mathbf{p} = [p_u, p_v], \tag{7}$$

$$\mathbf{A} = \frac{1}{4} B \left(u^2 + v^2 \right) [v, -u], \tag{8}$$

$$\mathbf{w} = [\dot{u}, \dot{v}] = \mathbf{p} - \mathbf{A}, \tag{9}$$

which allows us to rewrite the Hamiltonian in parabolic coordinates in a more compact form

$$h = \frac{1}{2} |\mathbf{p} - \mathbf{A}|^2 - E \left(u^2 + v^2 \right) - \frac{1}{2} F \left(u^4 - v^4 \right) - 2 \tag{10}$$

$$= \frac{1}{2} |\mathbf{w}|^2 - E \left(u^2 + v^2 \right) - \frac{1}{2} F \left(u^4 - v^4 \right) - 2. \tag{11}$$

The explicit dependence on the electric field strength F can be removed by rescaling the lengths by $F^{1/2}$, momenta by $F^{-1/4}$, time by $F^{3/4}$, magnetic field by $F^{-3/4}$, and the energy by $F^{-1/2}$. The resulting effect on Equation (10) is equivalent to setting $F = 1$. The system then depends on only two independent parameters: the magnetic field strength B and the electron energy E.

3. Construction of the Surface of Section

A global surface of section is easy to find for the case of hydrogen in *parallel* electric and magnetic fields. In that case the Hamiltonian can again be transformed using parabolic coordinates as in Section 2 yielding an effective Hamiltonian

$$h = \frac{1}{2} \left(p_u^2 + p_v^2 \right) - E \left(u^2 + v^2 \right) + \frac{1}{8} B^2 \left(u^4 v^2 + u^2 v^4 \right) - \frac{1}{2} \left(u^4 - v^4 \right) - 2. \tag{12}$$

In this case, the Hamiltonian can be decomposed into kinetic plus potential terms with no vector potential in the kinetic term. The shape of the potential guarantees that any trajectory launched from the u-axis will return to the u-axis crossing it transversely. Furthermore, every trajectory will intersect the u-axis an infinite number of times. We therefore can use the u-axis to construct the SOS; the SOS within the full four-dimensional phase space is the two-dimensional surface within the energy shell $h = 0$ that projects to the u-axis. This surface has two connected components: one for trajectories moving upward through the u-axis, and one for trajectories moving downward through the u-axis. We formally identify these two components via reflection symmetry about the u-axis. The SOS thus has canonical coordinates (u, p_u). The corresponding Poincaré map returns the value of the coordinates (u, p_u) each time a trajectory crosses the u-axis, going either upward or downward [23–25].

We can try the same approach in the crossed fields problem, defining a SOS with the u-axis. However, this approach fails due to the presence of the vector potential term in Equation (10). A trajectory launched tangent to the u-axis will curve away from it both forward and backward in time, due to the presence of the magnetic field; this launch point thus constitutes a tangential intersection with the u-axis (See Figure 1). Furthermore, the $\mathbf{E} \times \mathbf{B}$ drift causes some trajectories to never return to the u-axis (Figure 1). Hence this SOS definition would suffer both from tangencies and the failure to capture all possible trajectories.

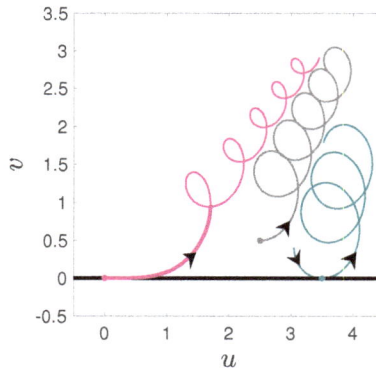

Figure 1. Three representative trajectories. Teal curve is the trajectory launched tangent to the u-axis away from the nucleus. Grey curve represents a trajectory launched off the u-axis away from the nucleus. Magenta trajectory is the trajectory launched away from the nucleus along the u-axis. The thick magenta curve represents the portion of the trajectory used to define the surface of section (SOS). ($B = 2$, $E = -1.1$, where B and E are the magnetic field strength and the energy in scaled coordinates.)

Returning to the parallel fields problem, we need to think more deeply about why the u-axis generates a good SOS in the parallel fields problem. For parallel fields, a trajectory launched tangent to the u-axis does not create a tangential intersection with the u-axis because it remains on the u-axis, that is, the u-axis is itself a trajectory of the system. This is true regardless of the magnitude of the momentum and whether the momentum points to the left or right. We adapt this idea for the crossed fields problem by choosing the trajectory that begins at the nucleus (located at $u = v = 0$) and is launched away from the nucleus along the u-axis to the right as shown in Figure 1. Locally at the nucleus this trajectory corresponds to the trajectory used to define the Poincaré return map in the parallel fields case. As the electron moves away from the nucleus, the electric and magnetic fields create a trajectory that resembles a tapered cycloid with the larger radius closer to the nucleus.

Notice that when projected onto the uv-space this trajectory self intersects. Nevertheless, we take the portion of the trajectory shown as bold in Figure 1 from the nucleus ($u = v = 0$) up to the point where the tangent to the trajectory points in the v direction in uv-space. (One could extend the trajectory up to the second time the trajectory passes through the first self-intersection point. However for ease of implementation, we are using a simpler criterion of cutting off the trajectory at the first vertical tangent line.) Next, we replicate this portion of the trajectory into the remaining quadrants in uv-space, by first reflecting through the origin and then reflecting about the horizontal axis (See Figure 2). Reflecting through the origin corresponds to launching a trajectory from the nucleus along the u-axis to the left. On the other hand, reflecting about the u-axis is equivalent to time reversal giving two trajectories that move toward the nucleus. In Figure 2, we show portions of the trajectories launched away from the nucleus in blue, and the time-reversed partners in red. We call these the SOS curves. The arrows denote the direction of time along the trajectory. Thus, a trajectory comes in towards the nucleus from the upper left and continues smoothly into the trajectory in the upper right.

Similarly a trajectory comes in from the lower right and exits on the lower left. These two trajectories are analogous to trajectories moving either left or right along the *u*-axis in the parallel fields case. In the parallel fields case these trajectories lie on top of one another, whereas here the vector potential splits them into separate curves.

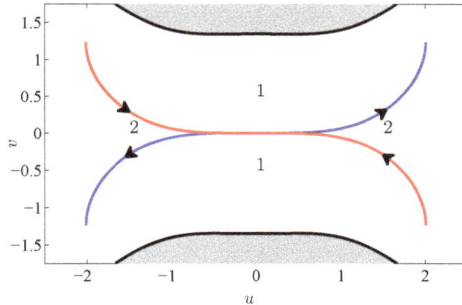

Figure 2. Trajectories defining the SOS ($B = 1.5$, $E = -0.2$). The thick black line represents the boundary between the energetically allowed (white) and forbidden (shaded) regions. Crossing either red or blue curves from region labeled 1 to region labeled 2 defines the SOS. Trajectories that cross from region 2 to region 1 are not part of the SOS.

The SOS is now defined by recording the electron's position and momentum whenever an electron trajectory intersects either the red or the blue curves while moving from the region labeled 1 to the region labeled 2. We do not consider crossings from region 2 to region 1 as part of the SOS. Thus we only consider upward propagating trajectories on the lower right and downward propagating trajectories on the upper right. For the parallel fields problem these two branches were identified, but this is no longer possible in the crossed fields case due to the broken time-reversal symmetry. Nevertheless, the picture can be simplified by taking advantage of another symmetry. There remains a reflection symmetry through the origin in which the upper left branch is identified with the lower right branch, and the lower left branch is identified with the upper right branch. We therefore compose the SOS from just the two branches on the right.

This definition of the SOS is free of tangencies for any trajectory approaching from region 1. This is because the curve that defines the SOS is a portion of a physical trajectory. To show this, suppose a trajectory does intersect the SOS transversely at an isolated point $\mathbf{d_0} = [u, v]$. Then the direction of the velocity $\mathbf{w} = [\dot{u}, \dot{v}]$ is determined up to a minus sign at r. Furthermore, since the trajectory has a fixed energy E, the magnitude of the velocity at $\mathbf{d_0}$ is determined by setting Equation (11) equal to zero and solving for $|\mathbf{w}|$. Thus, the velocity vector $\mathbf{w}$ is determined up to a minus sign. If the direction of $\mathbf{w}$ coincides with the direction of the trajectory used to define the SOS, then by the uniqueness of the ODE solution, the two trajectories are the same and the intersection $\mathbf{d_0}$ is not an isolated point. Hence, we take $\mathbf{w}$ to point in the opposite direction. In this case, the trajectory will curve in the opposite direction as the SOS curve, as shown in Figure 3. Therefore it intersects the SOS curve coming from region 2. Thus, no trajectory coming from region 1 can intersect the SOS tangentially.

We next introduce canonical coordinates in two-dimensional SOS. We define the position variable a to be the Euclidean length along the SOS trajectory as measured from the nucleus. Positive a parametrizes the curve in the upper right quadrant, and negative a parametrizes the curve in the lower right quadrant. The momentum coordinate p_a is defined by projecting the momentum (p_u, p_v) onto the tangent line to the SOS trajectory.

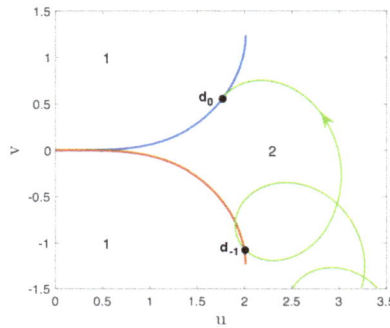

Figure 3. Green curve represents the trajectory which maps the point $\mathbf{d_{-1}}$ on the portion of the red SOS curve to the tangency $\mathbf{d_0}$ on the blue portion of the SOS curve.

We now define the Poincaré return map $(a, p_a) \mapsto (a', p_a')$ in the following way. Given (a, p_a) we begin a trajectory at position a along the SOS trajectory and with the tangential momentum given by p_a. We determine $|\mathbf{w}|$ by setting Equation (11) equal to zero. Please note that the tangential component of velocity is given by $\mathbf{w} \cdot \hat{\mathbf{t}} = p_a - \mathbf{A} \cdot \hat{\mathbf{t}}$, where $\hat{\mathbf{t}}$ is the unit tangent pointing forward along the SOS trajectory (Equation (9)). Thus, the perpendicular component of velocity is given by $w_\perp = \pm \sqrt{|\mathbf{w}|^2 - \left(p_a - \mathbf{A} \cdot \hat{\mathbf{t}}\right)^2}$. The sign of $w_\perp$ is determined by requiring the normal component of $\mathbf{w}$ to point from region 1 to region 2 in Figure 2. This uniquely determines $\mathbf{w}$, which uniquely determines $\mathbf{p}$ via Equation (9). Now that an initial point (u, v, p_u, p_v) in the full phase space has been determined, we evolve Hamilton's Equations (6) forward until the trajectory intersects one of the four branches in Figure 2 traveling from region 1 into region 2. If the intersection is with one of the two left branches, we reflect the point and its momentum through the origin, i.e., $(u, v, p_u, p_v) \mapsto -(u, v, p_u, p_v)$. At this point we record the coordinates (a', p_a') on the SOS. Please note that some trajectories will never return to the SOS since the SOS is not globally defined. Such trajectories are easy to identify because they spiral outward toward infinity, escaping the nucleus. See Figure 1.

A set of sample SOS plots is shown in Figure 4. One can notice the presence of periodic orbits and stable islands immersed in the chaotic sea.

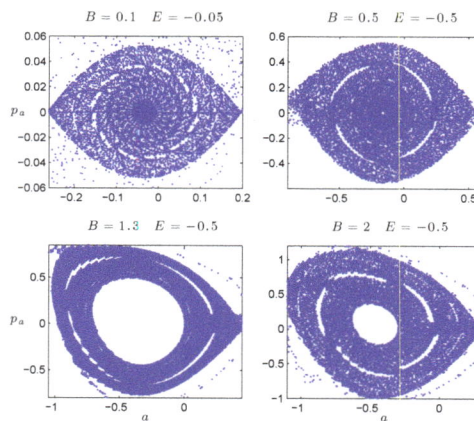

Figure 4. A sample of SOS plots for different values of the magnetic field strength and energy. The values of the magnetic field B and the energy of the ensemble E are marked on top of the figures. Electric field strength F is set to 1 in all four figures.

4. Visualizing a Periodic Orbit and Its Tangle

The ionization dynamics of this system is governed by a symmetric pair of periodic orbits in the left and right saddle regions of the potential in uv-space. (See Figure 5) In the original xy-coordinates these two periodic orbits are identified into a single orbit near the Stark saddle. In uv-space this orbit intersects the SOS curves in four places, but only two of those places intersect the SOS because the orbit must move from region 1 to region 2 as marked with black dots in Figure 6. These two intersection points form a period-2 orbit of the Poincaré map. This period-2 orbit is hyperbolic. Figure 7 shows a summary of the main results of this paper. The hyperbolic period-2 point has stable and unstable manifolds, shown in red and blue respectively, which intersect to form a heteroclinic tangle. Importantly the SOS for the parameters shown in Figure 7 is large enough to include not only the period-2 orbit but also the entire resonance zone defined by the tangle. The resonance zone is the domain bounded by the segments of the stable manifold (red) connecting z_R to p_0 and z_L to p_1 and by the segments of the unstable manifold (blue) connecting z_R to p_1 and z_L to p_0.

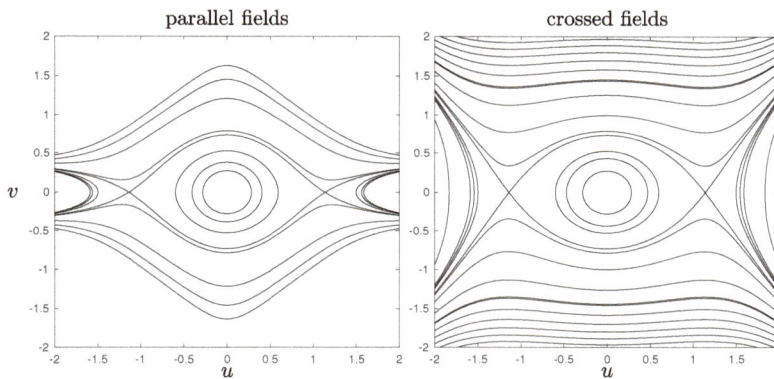

Figure 5. Left figure shows the potential for the parallel fields case, and the right figure shows the potential for the crossed fields case. ($E = -1.3$, and $B = 4.5$.)

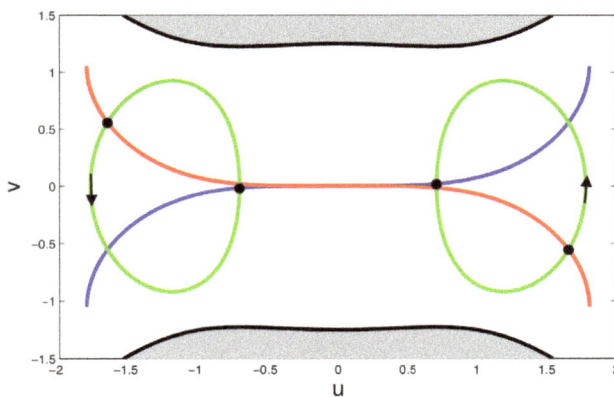

Figure 6. Symmetric unstable periodic orbits are shown in green. Two pairs of black dots mark the intersections of the periodic orbit with the SOS. The thick black line represents the boundary between the energetically allowed (white) and forbidden (shaded) regions.

The heteroclinic tangle defines regions in phase space called lobes, which fall into two categories: those governing the escape from the resonance zone, labeled E_k, and those governing the capture into the resonance zone, labeled C_k. (See Figure 7.) The escape lobe E_0, which is inside the resonance zone, maps to the E_1 lobe, which is outside the resonance zone. Similarly, the capture lobe C_{-1}, which is outside the resonance zone, maps to the C_0 lobe, which is inside. These lobes define a phase space turnstile which governs the ionization process [6].

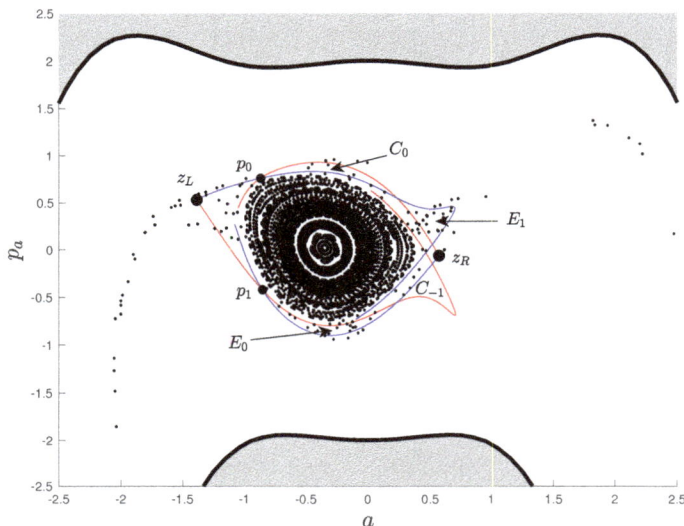

Figure 7. Surface of Section for $B = 1.5$, $E = -0.5$. Two big black dots represent the unstable period two orbit. The red line is the portion of the stable manifold, and the blue line is the portion of the unstable manifold. Thick black line defines the boundary between the energetically allowed (white) and forbidden (shaded) regions.

We now return to discuss the issue of tangencies and discontinuities in the Poincaré map. As mentioned before, there are no trajectories that intersect the SOS tangentially coming from region 1. However, in general there may be trajectories that map from the SOS to a tangential intersection with the SOS coming from region 2. See Figure 3. At any point on the SOS, i.e., at any value of the a coordinate, these tangential intersections correspond to the minimum value of momentum p_a. That is, the tangential intersections form the lower black boundary of the physically allowed domain in Figure 7. Thus, the Poincaré map will be continuous on any topological disk D that maps forward to a domain that does not intersect the lower boundary of the physically allowed region. It is enough to check just the boundary of the domain D. That is, if the boundary of a domain D maps forward to a closed curve that does not intersect the lower boundary of the physically allowed region, then the Poincaré map is continuous over the entire domain D. The resonance zone of the tangle in Figure 7 satisfies this property, and thus the Poincaré map is continuous over the entire resonance zone. The same argument applies to the inverse Poincaré map.

To complete the picture, we have also included additional orbits (represented by black dots) that are contained inside the heteroclinic tangle. One can observe a similar prominent chain of stable islands contained inside this heteroclinic tangle as the one shown in Figure 4. It is important to note that it is the heteroclinic tangle attached to the periodic orbit furthest away from the nucleus that governs the ionization process. Any additional heteroclinic tangles that might exist in the system would be contained inside this outermost tangle.

As we have mentioned earlier, this SOS is not defined globally. For any given set of parameter values, there are always trajectories that do not return to the SOS. For some set of parameter values, the SOS is large enough to encompass the whole of the resonance zone of the heteroclinic tangle. However, this does not hold for all parameter values but only for a certain range of magnetic field strength and energy. Two opposing effects determine the range of validity: the Euclidean distance from the nucleus where the trajectory has its first vertical tangent when projected to the uv-space, and the position of the unstable periodic orbit. For a fixed value of electron energy increasing the magnetic field decreases the Larmor radius of the SOS trajectory we use to define the Poincaré SOS and hence decreases the range of the a coordinate in the local SOS. At the same time the unstable period-two orbit moves away from the nucleus, and hence it its a coordinate increases, until it finally "slips off" the SOS. Figure 8 shows the B-dependance of the position of the periodic orbit and the edge of the SOS definition. For a critical value of the magnetic field strength, the periodic orbit falls off the surface of section, and this ceases to be a useful local SOS for representing the heteroclinic tangle structure. This happens at the value of B for which the blue and the green curves meet in Figure 8. (For $E = -0.5$ this failure occurs at $B = 1.91$.)

Since the portion of phase space that has been promoted from bound to ionized via a turnstile never returns to the bound state, we argue that the local SOS is sufficient if one is interested in studying the chaotic ionization mechanism in this system.

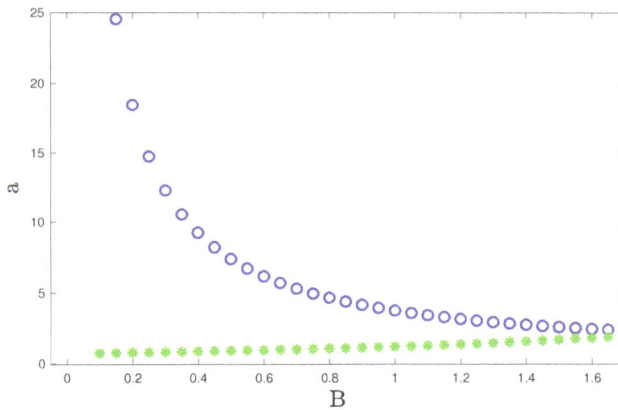

Figure 8. The blue dots represent the value of a at the edge of the surface of section, and the green stars represent the largest of the two absolute a values for the periodic orbit. ($E = -0.5$)

5. Conclusions

We have provided a prescription for defining a local surface of section for a nonlinear system whose global surface of section is not known and may not exist. This local surface of section allows us to capture all the main hallmarks of a chaotic system: an unstable periodic orbit, the most important pieces of the heteroclinic tangle attached to the unstable periodic orbit, the phase space turnstile that leads to ionization, as well as stable islands embedded in the chaotic sea. These structures can then be used for future applications such as the construction of the symbolic dynamics using the homotopic lobe dynamics approach, computation of topological entropy, periodic orbit computations of escape rates and spectral oscillations in the density of states [3,19,20]. One of the reasons why this technique works, which may be relevant for other applications, is that defining the SOS using an orbit of the system eliminates tangencies from trajectories propagating in the same direction as the SOS trajectory. Even though this approach fails for large magnetic field values, we believe it still offers valuable insights into studying types of problems where one is unable to define a global surface of section.

It would be interesting to see if this approach works in the circular restricted gravitational three-body problem as well.

Author Contributions: Conceptualization, K.B. and K.M.; Methodology, K.B. and K.M.; Software, K.B. and K.M., Validation, K.B. and K.M.; Formal Analysis, K.B. and K.M.; Writing-Original Draft Preparation, K.B. and K.M.; Writing-Review and Editing, K.B. and K.M.; Visualization, K.B. and K.M.; Funding Acquisition, K.M.

Funding: This work was supported in part by NSF grants PHY-0748828 and PHY-1408127. This material is based in part upon work supported by the National Science Foundation under Grant No. DMS-1440140 while the second author was in residence at the Mathematical Sciences Research Institute in Berkeley, California, during the Fall 2018 semester.

Conflicts of Interest: The authors declare no conflict of interest.

References

1. Strogatz, S. *Nonlinear Dynamics and Chaos: With Applications to Physics, Biology, Chemistry, and Engineering*; CRC Press: Boca Raton, FL, USA, 2018.
2. Meiss, J. *Differential Dynamical Systems, Revised Edition*; Mathematical Modeling and Computation, Society for Industrial and Applied Mathematics: Philadelphia, PA, USA, 2017.
3. Cvitanović, P.; Artuso, R.; Mainieri, R.; Tanner, G.; Vattay, G. *Chaos: Classical and Quantum*; Niels Bohr Inst.: Copenhagen, Denmark, 2016.
4. Wiggins, S. *Chaotic Transport in Dynamical Systems*; Springer: New York, NY, USA, 1992.
5. Rom-Kedar, V. Transport rates of a class of two-dimensional maps and flows. *Physica D* **1990**, *43*, 229–268. [CrossRef]
6. Mackay, R.; Meiss, J.; Percival, I. Transport in Hamiltonian systems. *Physica D* **1984**, *13*, 55–81. [CrossRef]
7. Jung, C.; Emmanouilidou, A. Construction of a natural partition of incomplete horseshoes. *Chaos* **2005**, *15*, 023101. [CrossRef] [PubMed]
8. Collins, P. Dynamics forced by surface trellises. In *Contemporary Mathematics: Geometry and Topology in Dynamics*; American Mathematical Society: Providence, RI, USA, 1999; pp. 65–86.
9. Collins, P. Symbolic Dynamics from homoclinic tangles. *Int. J. Bifurc. Chaos* **2002**, *12*, 605–617. [CrossRef]
10. Collins, P. Forcing Relations for Homoclinic Orbits of the Smale Horseshoe Map. *Exp. Math.* **2005**, *14*, 75–86. [CrossRef]
11. Burke, K.; Mitchell, K.A. Chaotic ionization of a Rydberg atom subjected to alternating kicks—The role of phase space turnstiles. *Phys. Rev. A* **2009**, *80*, 033416. [CrossRef]
12. Cushman, R.; Sadovskií, D. Monodromy in the hydrogen atom in crossed fields. *Physica D* **2000**, *142*, 166–196. [CrossRef]
13. Gekle, S.; Main, J.; Bartsch, T.; Uzer, T. Hydrogen atom in crossed electric and magnetic fields: Phase space topology and torus quantization via periodic orbits. *Phys. Rev. A* **2007**, *75*, 023406. [CrossRef]
14. Jaffé, C.; Farrelly, D.; Uzer, T. Transition State Theory without Time-Reversal Symmetry: Chaotic Ionization of the Hydrogen Atom. *Phys. Rev. Lett.* **2000**, *84*, 610–613. [CrossRef] [PubMed]
15. Von Milczewski, J.; Uzer, T. Chaos and order in crossed fields. *Phys. Rev. E* **1997**, *55*, 6540–6551. [CrossRef]
16. Schleif, C.R.; Delos, J.B. Monodromy and the structure of the energy spectrum of hydrogen in near perpendicular electric and magnetic fields. *Phys. Rev. A* **2007**, *76*, 013404. [CrossRef]
17. Wiebusch, G.; Main, J.; Krüger, K.; Rottke, H.; Holle, A.; Welge, K.H. Hydrogen atom in crossed magnetic and electric fields. *Phys. Rev. Lett.* **1989**, *62*, 2821–2824. [CrossRef] [PubMed]
18. Raithel, G.; Fauth, M.; Walther, H. Atoms in strong crossed electric and magnetic fields: Evidence for states with large electric-dipole moments, *Phys. Rev. A* **1993**, *47*, 419–440. [CrossRef]
19. Du, M.L.; Delos, J.B. Effect of closed classical orbits on quantum spectra: Ionization of atoms in a magnetic field. I. Physical picture and calculations. *Phys. Rev. A* **1988**, *38*, 1896–1912. [CrossRef]
20. Gao, J.; Delos, J.B.; Baruch, M. Closed-orbit theory of oscillations in atomic photoabsorption cross sections in a strong electric field. I. Comparison between theory and experiments on hydrogen and sodium above threshold. *Phys. Rev. A* **1992**, *46*, 1449–1454. [CrossRef] [PubMed]
21. Wang, D.M.; Delos, J.B. Organization and bifurcation of planar closed orbits of an atomic electron in crossed fields. *Phys. Rev. A* **2001**, *63*, 043409. [CrossRef]

22. Flöthmann, E.; Welge, K.H. Crossed-field hydrogen atom and the three-body Sun-Earth-Moon problem. *Phys. Rev. A* **1996**, *54*, 1884–1888. [CrossRef] [PubMed]
23. Mitchell, K.A.; Handley, J.P.; Tighe, B.; Flower, A.; Delos, J.B. Analysis of chaos-induced pulse trains in the ionization of hydrogen. *Phys. Rev. A* **2004**, *70*, 043407. [CrossRef]
24. Mitchell, K.A.; Handley, J.P.; Tighe, B.; Flower, A.; Delos, J.B. Chaos-Induced Pulse Trains in the Ionization of Hydrogen. *Phys. Rev. Lett.* **2004**, *92*, 073001. [CrossRef] [PubMed]
25. Mitchell, K.A.; Delos, J.B. The structure of ionizing electron trajectories for hydrogen in parallel fields. *Physica D* **2007**, *229*, 9–21. [CrossRef]

mathematics

MDPI

Article

Stability Analysis of an Age-Structured SIR Epidemic Model with a Reduction Method to ODEs

Toshikazu Kuniya [ORCID]

Graduate School of System Informatics, Kobe University, 1-1 Rokkodai-cho, Nada-ku, Kobe 657-8501, Japan; tkuniya@port.kobe-u.ac.jp

Received: 21 July 2018; Accepted: 22 August 2018; Published: 23 August 2018

Abstract: In this paper, we are concerned with the asymptotic stability of the nontrivial endemic equilibrium of an age-structured susceptible-infective-recovered (SIR) epidemic model. For a special form of the disease transmission function, we perform the reduction of the model into a four-dimensional system of ordinary differential equations (ODEs). We show that the unique endemic equilibrium of the reduced system exists if the basic reproduction number for the original system is greater than unity. Furthermore, we perform the stability analysis of the endemic equilibrium and obtain a fourth-order characteristic equation. By using the Routh–Hurwitz criterion, we numerically show that the endemic equilibrium is asymptotically stable in some epidemiologically relevant parameter settings.

Keywords: SIR epidemic model; age structure; endemic equilibrium; stability; basic reproduction number

MSC: 34D20; 37N25; 92D30

1. Introduction

The mathematical modeling of epidemics in human populations has been studied for a long time [1]. In 1760, Bernoulli used a mathematical model of differential equations to discuss the benefit of smallpox inoculation [2]. In 1911, Ross claimed that malaria could be eradicated by reducing the number of mosquitoes, and constructed a mathematical model of differential equations to theoretically support his claim [3]. In 1927, Kermack and McKendrick constructed the first susceptible-infective-recovered (SIR) epidemic model, in which the total population is divided into three classes called susceptible, infective, and recovered [4]. Since their work, the theory of various epidemic models such as a susceptible-infective-susceptible (SIS) epidemic model [5], a susceptible-exposed-infective-recovered (SEIR)epidemic model [6], and a susceptible-infective-recovered-susceptible (SIRS) epidemic model [7] with various structures such as the age structure [8,9], the space structure [10], and the network structure [11] has been developed from both mathematical and epidemiological points of view.

Epidemiologically, the basic reproduction number $\mathcal{R}_0$ for an infectious disease is defined by the expected number of secondary cases produced by a typical infective individual in a completely susceptible population ([9], Chapter 5). Mathematically, $\mathcal{R}_0$ is defined by the spectral radius of a linear operator called the next-generation operator [12], and it determines the complete global dynamics of each equilibrium for some basic epidemic models: if $\mathcal{R}_0 < 1$, then the trivial disease-free equilibrium is globally asymptotically stable, whereas if $\mathcal{R}_0 > 1$, then the nontrivial endemic equilibrium is globally asymptotically stable [13]. However, for some epidemic models, the endemic equilibrium can be stable even if $\mathcal{R}_0 < 1$ due to the backward bifurcation [14], and it can be unstable even if $\mathcal{R}_0 > 1$, which leads to a periodic solution due to the Hopf bifurcation [15].

In [16], some conjectures on the threshold property of $\mathcal{R}_0$ for an age-structured SIR epidemic model were proposed, and they were proved in [17]: if $\mathcal{R}_0 < 1$, then the disease-free equilibrium is globally asymptotically stable and no endemic equilibrium exists, whereas if $\mathcal{R}_0 > 1$, then the endemic equilibrium uniquely exists and it is locally asymptotically stable under some additional conditions.

However, in general, it is known that the endemic equilibrium cannot always be unique and stable for $\mathcal{R}_0 > 1$. Several authors have studied some special cases where the endemic equilibrium is unstable, and periodic solutions occur through the Hopf bifurcation for $\mathcal{R}_0 > 1$ [18–21]. From the viewpoint of application, it is important to investigate when the endemic equilibrium of an age-structured SIR epidemic model is stable and when it is not, as the age distribution of infective individuals in such a model should be stable if one tries to estimate the basic reproduction number $\mathcal{R}_0$ for an endemic disease by using statistical data that exhibit an almost unchanged age distribution of infective individuals. The purpose of this study is to obtain a new condition under which the endemic equilibrium of an age-structured SIR epidemic model is (locally) asymptotically stable.

Age-structured SIR epidemic models as stated above are systems of partial differential equations (PDEs), and hence the stability analysis of them often requires a relatively difficult method such as the spectral theory of positive operators ([17], §5). In this paper, we make some assumptions on the parameters of an age-structured SIR epidemic model, under which we can perform the reduction of the model into a four-dimensional system of ordinary differential equations (ODEs). We can then apply the standard method of characteristic equations for the stability analysis of the endemic equilibrium.

This paper is organized as follows. In Section 2, we formulate an age-structured SIR epidemic model, and perform the reduction of it into a four-dimensional system of ODEs. In Section 3, we prove that the reduced system has a unique endemic equilibrium if the basic reproduction number $\mathcal{R}_0$ for the original system is greater than unity. Moreover, we investigate the asymptotic stability of the endemic equilibrium, and obtain a fourth-order characteristic equation. As its coefficients have quite complex forms, we only prove their positivity, and numerically show by using the Routh–Hurwitz criterion that the endemic equilibrium is asymptotically stable in some epidemiologically relevant parameter settings in Section 4. In the parameter settings, the essential supremum of the demographic mortality rate is determined based on a dataset for Japan in 2015, and the recovery rate γ and the basic reproduction number $\mathcal{R}_0$ are varied for the cases of an influenza-like disease ($\gamma = 52$ and $2 \leq \mathcal{R}_0 \leq 3$), a chlamydia-like sexually transmitted disease ($\gamma = 1$ and $1 < \mathcal{R}_0 \leq 1.5$), and a wider range of realistic values of them ($1/50 \leq \gamma \leq 365$ and $1 < \mathcal{R}_0 \leq 50$). Finally, Section 5 is devoted to the discussion.

2. Reduction of an Age-Structured SIR Epidemic Model into ODEs

We first formulate an age-structured SIR epidemic model. Let $S(t, a)$, $I(t, a)$, and $R(t, a)$ denote the susceptible, infective, and recovered populations of age $a \geq 0$ at time $t \geq 0$, respectively. Let $b > 0$ denote the birth rate, let $\mu(\cdot) \in L^{\infty}_+(0, +\infty)$ denote the age-specific mortality rate such that $\int_0^{+\infty} \mu(a)\mathrm{d}a = +\infty$, and let $\gamma > 0$ denote the recovery rate. As in [18], we focus on the case where the disease transmission function is only dependent on the age of infective individuals: $\kappa = \kappa(\cdot) \in L^{\infty}_+(0, +\infty)$. In this case, the age-structured SIR epidemic model is formulated as follows:

$$\begin{cases} \left(\dfrac{\partial}{\partial t} + \dfrac{\partial}{\partial a}\right) S(t, a) = -S(t, a) \displaystyle\int_0^{+\infty} \kappa(a)I(t, a)\mathrm{d}a - \mu(a)S(t, a), & t > 0,\ a > 0, \\[2mm] \left(\dfrac{\partial}{\partial t} + \dfrac{\partial}{\partial a}\right) I(t, a) = S(t, a) \displaystyle\int_0^{+\infty} \kappa(a)I(t, a)\mathrm{d}a - [\mu(a) + \gamma]\, I(t, a), & t > 0,\ a > 0, \\[2mm] \left(\dfrac{\partial}{\partial t} + \dfrac{\partial}{\partial a}\right) R(t, a) = \gamma I(t, a) - \mu(a)R(t, a), & t > 0,\ a > 0, \\[2mm] S(t, 0) = b, \quad I(t, 0) = R(t, 0) = 0, & t > 0, \\[2mm] S(0, a) = S_0(a), \quad I(0, a) = I_0(a), \quad R(0, a) = R_0(a), & a \geq 0, \end{cases} \tag{1}$$

where $(S_0(\cdot), I_0(\cdot), R_0(\cdot)) \in L^1_+(0, +\infty) \times L^1_+(0, +\infty) \times L^1_+(0, +\infty)$ denotes the initial age distributions of each population. It is easy to see that (1) has the demographic steady state $P^*(a) = b e^{-\int_0^a \mu(\sigma)\mathrm{d}\sigma}$, $a \geq 0$. Let

$$s(t, a) = \frac{S(t, a)}{P^*(a)}, \quad i(t, a) = \frac{I(t, a)}{P^*(a)}, \quad r(t, a) = \frac{R(t, a)}{P^*(a)}, \quad t \geq 0,\ a \geq 0.$$

We then can normalize (1) as follows.

$$
\begin{cases}
\left(\dfrac{\partial}{\partial t} + \dfrac{\partial}{\partial a}\right) s(t,a) = -s(t,a)\displaystyle\int_0^{+\infty} \kappa(a)P^*(a)i(t,a)\mathrm{d}a, & t > 0,\ a > 0, \\[2mm]
\left(\dfrac{\partial}{\partial t} + \dfrac{\partial}{\partial a}\right) i(t,a) = s(t,a)\displaystyle\int_0^{+\infty} \kappa(a)P^*(a)i(t,a)\mathrm{d}a - \gamma i(t,a), & t > 0,\ a > 0, \\[2mm]
\left(\dfrac{\partial}{\partial t} + \dfrac{\partial}{\partial a}\right) r(t,a) = \gamma i(t,a), & t > 0,\ a > 0, \\[2mm]
s(t,0) = 1, \quad i(t,0) = r(t,0) = 0, & t > 0, \\[2mm]
s(0,a) = \dfrac{S_0(a)}{P^*(a)}, \quad i(0,a) = \dfrac{I_0(a)}{P^*(a)}, \quad r(0,a) = \dfrac{R_0(a)}{P^*(a)}, & a \geq 0.
\end{cases}
\tag{2}
$$

As shown in [18], for a specific form of $\kappa(\cdot)$ such that $\kappa(\cdot)P^*(\cdot)$ is sufficiently concentrated in one particular age class, the endemic equilibrium of (2) can be destabilized even if it uniquely exists. Thus, our interest in this paper is when it is stable. In this paper, we assume that $\kappa(\cdot)$ has the following form:

$$
\kappa(a) = \frac{\beta}{b} a e^{-ka} e^{\int_0^a \mu(\sigma)\mathrm{d}\sigma}, \quad \beta > 0,\ k > \mu^\infty = \operatorname*{ess.sup}_{a \geq 0} \mu(a) > 0, \quad a \geq 0.
\tag{3}
$$

Note that $\kappa(\cdot) \in L^\infty_+(0, +\infty)$ is satisfied under this assumption. In this case, the equations of s and i in (2) can be rewritten as follows.

$$
\begin{cases}
\left(\dfrac{\partial}{\partial t} + \dfrac{\partial}{\partial a}\right) s(t,a) = -s(t,a)\beta\displaystyle\int_0^{+\infty} a e^{-ka}i(t,a)\mathrm{d}a, & t > 0,\ a > 0, \\[2mm]
\left(\dfrac{\partial}{\partial t} + \dfrac{\partial}{\partial a}\right) i(t,a) = s(t,a)\beta\displaystyle\int_0^{+\infty} a e^{-ka}i(t,a)\mathrm{d}a - \gamma i(t,a), & t > 0,\ a > 0, \\[2mm]
s(t,0) = 1, \quad i(t,0) = 0, & t > 0, \\[2mm]
s(0,a) = \dfrac{S_0(a)}{P^*(a)}, \quad i(0,a) = \dfrac{I_0(a)}{P^*(a)}, & a \geq 0.
\end{cases}
\tag{4}
$$

Note that we can omit the equation of r, as it does not affect the dynamics of (4). As in [17], without loss of generality, we can assume that $0 \leq s(t,a) \leq 1$ and $0 \leq i(t,a) \leq 1$ for all $t \geq 0$ and $a \geq 0$.

As seen in [17,18], the stability analysis of age-structured PDE systems such as (4) requires complex calculation. In this paper, we perform the reduction of (4) into a four-dimensional system of ODEs. Let

$$
X(t) = \int_0^{+\infty} e^{-ka}s(t,a)\mathrm{d}a, \quad Y(t) = \int_0^{+\infty} e^{-ka}i(t,a)\mathrm{d}a,
$$

$$
L(t) = \beta\int_0^{+\infty} a e^{-ka}s(t,a)\mathrm{d}a, \quad \Lambda(t) = \beta\int_0^{+\infty} a e^{-ka}i(t,a)\mathrm{d}a, \quad t \geq 0.
$$

Note that these variables have no specific epidemiological implications except $\Lambda(t)$, which implies the force of infection at time $t \geq 0$. By differentiating $X(\cdot)$, we have

$$
\begin{aligned}
\frac{\mathrm{d}X(t)}{\mathrm{d}t} &= \int_0^{+\infty} e^{-ka}\frac{\partial s(t,a)}{\partial t}\mathrm{d}a = \int_0^{+\infty} e^{-ka}\left[-\frac{\partial s(t,a)}{\partial a} - s(t,a)\Lambda(t)\right]\mathrm{d}a \\[2mm]
&= \left[-e^{-ka}s(t,a)\right]_0^{+\infty} - k\int_0^{+\infty} e^{-ka}s(t,a)\mathrm{d}a - X(t)\Lambda(t) \\[2mm]
&= 1 - [k + \Lambda(t)]\,X(t), \quad t > 0.
\end{aligned}
\tag{5}
$$

Similarly, we have

$$
\begin{aligned}
\frac{dY(t)}{dt} &= \int_0^{+\infty} e^{-ka}\frac{\partial i(t,a)}{\partial t}da = \int_0^{+\infty} e^{-ka}\left[-\frac{\partial i(t,a)}{\partial a} + s(t,a)\Lambda(t) - \gamma i(t,a)\right]da \\
&= \left[-e^{-ka}i(t,a)\right]_0^{+\infty} - k\int_0^{+\infty} e^{-ka}i(t,a)da + X(t)\Lambda(t) - \gamma Y(t) \quad (6)\\
&= X(t)\Lambda(t) - (k+\gamma)\,Y(t), \quad t > 0,
\end{aligned}
$$

$$
\begin{aligned}
\frac{dL(t)}{dt} &= \beta\int_0^{+\infty} ae^{-ka}\frac{\partial s(t,a)}{\partial t}da = \beta\int_0^{+\infty} ae^{-ka}\left[-\frac{\partial s(t,a)}{\partial a} - s(t,a)\Lambda(t)\right]da \\
&= \beta\left[-ae^{-ka}s(t,a)\right]_0^{+\infty} + \beta\int_0^{+\infty} e^{-ka}s(t,a)da - k\int_0^{+\infty} ae^{-ka}s(t,a)da - L(t)\Lambda(t) \quad (7)\\
&= \beta X(t) - [k+\Lambda(t)]\,L(t), \quad t > 0,
\end{aligned}
$$

and

$$
\begin{aligned}
\frac{d\Lambda(t)}{dt} &= \beta\int_0^{+\infty} ae^{-ka}\frac{\partial i(t,a)}{\partial t}da = \beta\int_0^{+\infty} ae^{-ka}\left[-\frac{\partial i(t,a)}{\partial a} + s(t,a)\Lambda(t) - \gamma i(t,a)\right]da \\
&= \beta\left[-ae^{-ka}i(t,a)\right]_0^{+\infty} + \beta\int_0^{+\infty} e^{-ka}i(t,a)da - k\int_0^{+\infty} ae^{-ka}i(t,a)da + L(t)\Lambda(t) - \gamma\Lambda(t) \quad (8)\\
&= \beta Y(t) + L(t)\Lambda(t) - (k+\gamma)\,\Lambda(t), \quad t > 0.
\end{aligned}
$$

Hence, combining (5)–(8), we obtain the following new four-dimensional system of ODEs:

$$
\begin{cases}
\dfrac{dX(t)}{dt} = 1 - [k+\Lambda(t)]\,X(t), & t > 0, \\[2mm]
\dfrac{dY(t)}{dt} = X(t)\Lambda(t) - (k+\gamma)\,Y(t), & t > 0, \\[2mm]
\dfrac{dL(t)}{dt} = \beta X(t) - [k+\Lambda(t)]\,L(t), & t > 0, \\[2mm]
\dfrac{d\Lambda(t)}{dt} = \beta Y(t) + L(t)\Lambda(t) - (k+\gamma)\,\Lambda(t), & t > 0, \\[2mm]
X(0) = X_0, \quad Y(0) = Y_0, \quad L(0) = L_0, \quad \Lambda(0) = \Lambda_0.
\end{cases}
\quad (9)
$$

Note that $(X_0, Y_0, L_0, \Lambda_0) \in \mathbb{R}_+^4$ since $(S_0(\cdot), I_0(\cdot)) \in L_+^1(0,+\infty) \times L_+^1(0,+\infty)$. In what follows, we perform the stability analysis of system (9).

3. Existence, Uniqueness, and Stability of the Endemic Equilibrium

Following the theory in [12], the basic reproduction number $\mathcal{R}_0$ for the original (normalized) system (4) can be calculated as follows.

$$
\begin{aligned}
\mathcal{R}_0 &= \beta\int_0^{+\infty} ae^{-ka}\int_0^a e^{-\gamma(a-\sigma)}d\sigma da = \frac{\beta}{\gamma}\int_0^{+\infty} ae^{-ka}\left(1 - e^{-\gamma a}\right)da \\
&= \frac{\beta}{\gamma}\left\{\left[-\frac{ae^{-ka}}{k}\right]_0^{+\infty} + \frac{1}{k}\int_0^{+\infty} e^{-ka}da - \left[-\frac{ae^{-(k+\gamma)a}}{k+\gamma}\right]_0^{+\infty} - \frac{1}{k+\gamma}\int_0^{+\infty} e^{-(k+\gamma)a}da\right\} \quad (10)\\
&= \frac{\beta}{\gamma}\left[\frac{1}{k^2} - \frac{1}{(k+\gamma)^2}\right] = \beta\frac{2k+\gamma}{k^2(k+\gamma)^2}.
\end{aligned}
$$

We now prove that system (9) has the unique endemic equilibrium if $\mathcal{R}_0 > 1$. Let $E^* : (X^*, Y^*, L^*, \Lambda^*) \in (\mathbb{R}_+ \setminus \{0\})^4$ denote the endemic equilibrium of system (9). The following equations are satisfied:

$$
\begin{cases}
1 - (k + \Lambda^*) X^* = 0, \\
X^* \Lambda^* - (k + \gamma) Y^* = 0, \\
\beta X^* - (k + \Lambda^*) L^* = 0, \\
\beta Y^* + L^* \Lambda^* - (k + \gamma) \Lambda^* = 0.
\end{cases}
\tag{11}
$$

We prove the following theorem.

Theorem 1. *Let $\mathcal{R}_0$ be defined by (10). If $\mathcal{R}_0 > 1$, then system (9) has the unique endemic equilibrium E^*.*

Proof. By rearranging the first three equations in (11), we have

$$
X^* = \frac{1}{k + \Lambda^*}, \quad Y^* = \frac{X^* \Lambda^*}{k + \gamma} = \frac{\Lambda^*}{(k + \gamma)(k + \Lambda^*)}, \quad L^* = \frac{\beta X^*}{k + \Lambda^*} = \frac{\beta}{(k + \Lambda^*)^2}.
\tag{12}
$$

By substituting the equations of Y^* and L^* into the last equation in (11), we have

$$
0 = \frac{\beta \Lambda^*}{(k + \gamma)(k + \Lambda^*)} + \frac{\beta \Lambda^*}{(k + \Lambda^*)^2} - (k + \gamma) \Lambda^*.
$$

Dividing both sides of this equation by Λ^* and rearranging the equation, we have

$$
1 = \frac{\beta}{k + \gamma} \left\{ \frac{1}{(k + \gamma)(k + \Lambda^*)} + \frac{1}{(k + \Lambda^*)^2} \right\}.
\tag{13}
$$

Let $F(\Lambda^*)$ be a function defined by the right-hand side of this equation. Since $F(\Lambda^*)$ is monotonically decreasing to 0 as $\Lambda^* \to +\infty$ and

$$
F(0) = \frac{\beta}{k + \gamma} \left\{ \frac{1}{(k + \gamma) k} + \frac{1}{k^2} \right\} = \beta \frac{2k + \gamma}{k^2 (k + \gamma)^2} = \mathcal{R}_0 > 1,
$$

there exists the unique positive root $\Lambda^* > 0$ of Equation (13). Substituting it into the three equations in (12), we obtain the unique endemic equilibrium E^*. This completes the proof. $\square$

To investigate the asymptotic stability of the endemic equilibrium E^*, we consider the following Jacobian matrix $\mathbf{J}_{E^*}$ around E^*:

$$
\mathbf{J}_{E^*} = \begin{pmatrix}
-(k + \Lambda^*) & 0 & 0 & -X^* \\
\Lambda^* & -(k + \gamma) & 0 & X^* \\
\beta & 0 & -(k + \Lambda^*) & -L^* \\
0 & \beta & \Lambda^* & L^* - (k + \gamma)
\end{pmatrix}.
$$

From (12), we have

$$
k + \Lambda^* = \frac{1}{X^*}, \quad L^* = \beta (X^*)^2.
\tag{14}
$$

Using (14), we derive the characteristic polynomial for E^* as follows:

$$|\lambda \mathbf{I} - \mathbf{J}_{E^*}| = \begin{vmatrix} \lambda + k + \Lambda^* & 0 & 0 & X^* \\ -\Lambda^* & \lambda + k + \gamma & 0 & -X^* \\ -\beta & 0 & \lambda + k + \Lambda^* & L^* \\ 0 & -\beta & -\Lambda^* & \lambda - L^* + k + \gamma \end{vmatrix}$$

$$= \begin{vmatrix} \lambda + k + \Lambda^* & 0 & 0 & X^* \\ \lambda + k & \lambda + k + \gamma & 0 & 0 \\ -\beta & 0 & \lambda + k + \Lambda^* & \beta(X^*)^2 \\ -\beta & -\beta & \lambda + k & \lambda + k + \gamma \end{vmatrix} \tag{15}$$

$$= \begin{vmatrix} \lambda + \dfrac{1}{X^*} & 0 & 0 & X^* \\ -\gamma & \lambda + k + \gamma & 0 & 0 \\ -\beta(X^*\lambda + 2) & 0 & \lambda + \dfrac{1}{X^*} & 0 \\ 0 & -\beta & \lambda + k & \lambda + k + \gamma \end{vmatrix}$$

$$= \left(\lambda + \dfrac{1}{X^*}\right)^2 (\lambda + k + \gamma)^2 - \beta X^* \left[-\gamma \left(\lambda + \dfrac{1}{X^*}\right) + (\lambda + k)(\lambda + k + \gamma)(X^*\lambda + 2)\right]$$

$$= \lambda^4 + a_3 \lambda^3 + a_2 \lambda^2 + a_1 \lambda + a_0,$$

where $\mathbf{I}$ denotes the identity matrix and

$$\begin{cases} a_3 = \dfrac{2}{X^*} + 2(k + \gamma) - \beta(X^*)^2, \\[2mm] a_2 = \dfrac{1}{(X^*)^2} + 4\dfrac{k+\gamma}{X^*} + (k+\gamma)^2 - \beta X^*\{2 + (2k+\gamma)X^*\}, \\[2mm] a_1 = \dfrac{2(k+\gamma)}{(X^*)^2} + \dfrac{2(k+\gamma)^2}{X^*} - \beta X^*\{-\gamma + k(k+\gamma)X^* + 2(2k+\gamma)\}, \\[2mm] a_0 = \dfrac{(k+\gamma)^2}{(X^*)^2} - \beta X^*\left\{-\dfrac{\gamma}{X^*} + 2k(k+\gamma)\right\}. \end{cases} \tag{16}$$

To apply the Routh–Hurwitz criterion, we prove the following proposition.

Proposition 1. *Let $\mathcal{R}_0$ and a_i ($i = 0, 1, 2, 3$) be defined by (10) and (16), respectively. If $\mathcal{R}_0 > 1$, then $a_i > 0$ for all $i = 0, 1, 2, 3$.*

Proof. By Theorem 1, the unique endemic equilibrium E^* exists. From (12) and (13), we have

$$k + \gamma = \dfrac{\beta X^*}{k + \gamma} + \beta(X^*)^2, \quad (k+\gamma)^2 = \beta X^* + (k+\gamma)\beta(X^*)^2, \quad 1 - kX^* = 1 - \dfrac{k}{k + \Lambda^*} > 0.$$

We then have

$$a_3 = \frac{2}{X^*} + 2\left\{\frac{\beta X^*}{k+\gamma} + \beta (X^*)^2\right\} - \beta (X^*)^2 = \frac{2}{X^*} + \frac{2\beta X^*}{k+\gamma} + \beta (X^*)^2 > 0,$$

$$a_2 = \frac{1}{(X^*)^2} + \frac{4}{X^*}\left\{\frac{\beta X^*}{k+\gamma} + \beta (X^*)^2\right\} + \beta X^* + (k+\gamma)\beta (X^*)^2 - \beta X^* \{2 + (2k+\gamma)X^*\}$$

$$= \frac{1}{(X^*)^2} + \frac{4\beta}{k+\gamma} + 3\beta X^* - k\beta (X^*)^2 = \frac{1}{(X^*)^2} + \frac{4\beta}{k+\gamma} + 2\beta X^* + (1 - kX^*)\beta X^* > 0,$$

$$a_1 = \frac{2\beta}{(k+\gamma) X^*} + 2\beta + 2\beta + 2 (k+\gamma)\beta X^* + \gamma \beta X^* - k (k+\gamma)\beta (X^*)^2 - 2k\beta X^* - 2 (k+\gamma)\beta X^*$$

$$= \frac{2\beta}{(k+\gamma) X^*} + 2\beta (1 - kX^*) + 2\beta + (k+\gamma)\beta X^* - k\beta X^* - k (k+\gamma)\beta (X^*)^2$$

$$= \frac{2\beta}{(k+\gamma) X^*} + 3\beta (1 - kX^*) + \beta + (k+\gamma)\beta X^* (1 - kX^*) > 0,$$

$$a_0 = \frac{\beta}{X^*} + (k+\gamma)\beta + \gamma\beta - 2k (k+\gamma)\beta X^* = \frac{\beta}{X^*} + 2 (k+\gamma)\beta - k\beta - 2k (k+\gamma)\beta X^*$$

$$= \frac{\beta}{X^*} (1 - kX^*) + 2 (k+\gamma)\beta (1 - kX^*) > 0.$$

This completes the proof. □

By Proposition 1, it follows from the Routh–Hurwitz criterion [15] (Proof of Theorem 3.1) that the endemic equilibrium E^* is asymptotically stable if and only if

$$\Delta = a_1 a_2 a_3 - a_0 a_3^2 - a_1^2 > 0. \tag{17}$$

However, it seems that quite a long calculation is needed to show (17) analytically. Instead, in the next section, we show (17) numerically in some epidemiologically relevant parameter settings.

4. Numerical Results

Let the unit time be 1 year. By the definition in (3), k should satisfy inequality $k > \mu^\infty = \text{ess.sup}_{a \geq 0} \mu(a) > 0$. For the sake of simplicity, we regard $a = 100$ (years old) as the maximum age of individuals. In a dataset available in [22], the mortality rate $\mu(a)$ is at most 0.39954 (at $a = 100$) for males in Japan, 2015. Hence, we fix $k = 0.4$. By (10), we can determine β for chosen $\mathcal{R}_0$ and γ as follows:

$$\beta = \frac{k^2 (k+\gamma)^2}{2k+\gamma} \mathcal{R}_0.$$

We first consider an influenza-like disease which has an infectious period of about 1 week (see [23]). Therefore, let $\gamma = 52$ so that the average infectious period is $1/\gamma = 1/52$ year = 1 week. Following the estimation result in [24], we vary the value of $\mathcal{R}_0$ from 2 to 3. In Figure 1a, we can confirm that criterion Δ defined in (17) is always positive, and hence, the endemic equilibrium E^* is asymptotically stable. In fact, in Figure 1b, the force of infection $\Lambda(t)$ converges to the positive equilibrium value $\Lambda^* = 0.2339 > 0$ as time evolves for $\mathcal{R}_0 = 2.5$.

We next consider a chlamydia-like sexually transmitted disease which has an infectious period of about 1 year (see [25]). Therefore, $\gamma = 1$ so that the average infectious period is $1/\gamma = 1$ year. Based on the estimation result in [25], we vary the value of $\mathcal{R}_0$ from 1 to 1.5. In Figure 2a, we can confirm that criterion Δ is always positive, and hence the endemic equilibrium E^* is asymptotically stable. In fact, in Figure 2b, the force of infection $\Lambda(t)$ converges to the positive equilibrium value $\Lambda^* = 0.0539 > 0$ as time evolves for $\mathcal{R}_0 = 1.25$.

Finally, we vary the values of γ from 1/50 to 365 (i.e., the average infectious period $1/\gamma$ is varied from 1/365 year = 1 day to 50 years) and $\mathcal{R}_0$ from 1 to 50. In Figure 3, we can confirm that criterion Δ

is always positive in this parameter region. Hence, we can conclude that the endemic equilibrium E^* is asymptotically stable for epidemiologically relevant values of γ and $\mathcal{R}_0$.

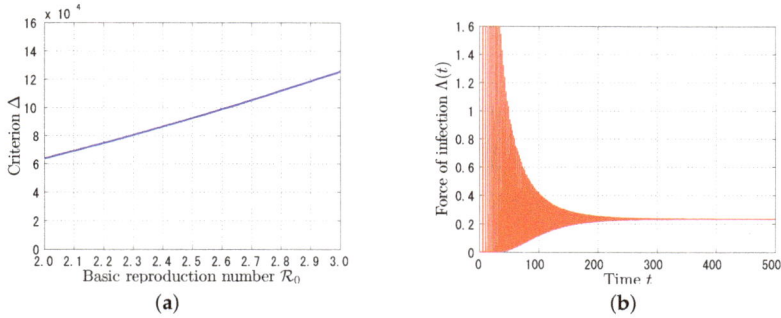

Figure 1. Numerical confirmation of condition (17) and asymptotic stability of the endemic equilibrium E^* for the case of an influenza-like disease: (**a**) Variation of Δ defined in (17) for $2 \leq \mathcal{R}_0 \leq 3$; (**b**) Time variation of $\Lambda(t)$ for $\mathcal{R}_0 = 2.5$.

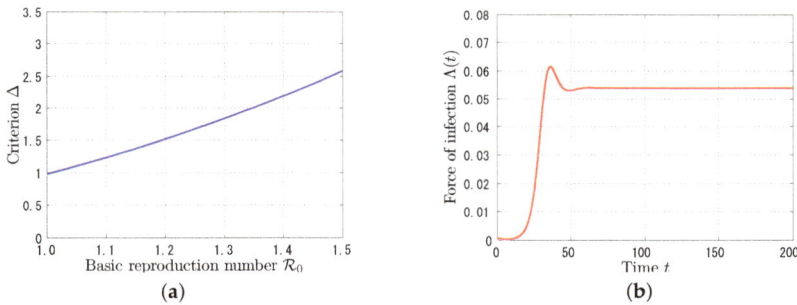

Figure 2. Numerical confirmation of condition (17) and asymptotic stability of the endemic equilibrium E^* for the case of a chlamydia-like sexually transmitted disease: (**a**) Variation of Δ defined in (17) for $1 < \mathcal{R}_0 \leq 1.5$; (**b**) Time variation of $\Lambda(t)$ for $\mathcal{R}_0 = 1.25$.

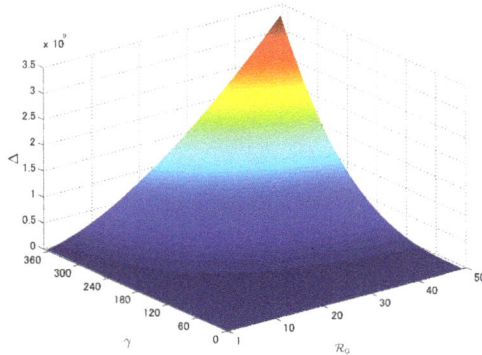

Figure 3. Numerical confirmation of condition (17) for realistic values of recovery rate $\gamma \in [1/50, 365]$ and the basic reproduction number $\mathcal{R}_0 \in (1, 50]$.

5. Discussion

In this paper, we formulated an age-structured SIR epidemic model and performed its reduction into a four-dimensional system of ODEs under an additional assumption on the disease transmission function $\kappa(\cdot)$. We proved that if the basic reproduction number $\mathcal{R}_0$ is greater than 1, then the system has the unique endemic equilibrium E^* (Theorem 1). Moreover, we obtained a fourth-order characteristic equation, and proved that all of its coefficients are positive (Proposition 1). By the Routh–Hurwitz criterion, the endemic equilibrium E^* is asymptotically stable if and only if $\Delta > 0$. As it seems difficult to show $\Delta > 0$ analytically, we showed it numerically in Section 4 for some epidemiologically relevant parameters. We showed that $\Delta > 0$ holds for parameters for an influenza-like disease ($\gamma = 52$ and $2 \leq \mathcal{R}_0 \leq 3$) and a chlamydia-like disease ($\gamma = 1$ and $1 < \mathcal{R}_0 \leq 1.5$). Furthermore, we showed that $\Delta > 0$ holds for a wider region of epidemiologically relevant parameters $\gamma \in [1/50, 365]$ and $\mathcal{R}_0 \in (1, 50]$.

The results in this paper contribute to enlarge the stability region of the endemic equilibrium E^* for $\mathcal{R}_0 > 1$ to the set where we can perform our reduction method of the PDEs system into the ODEs system. Epidemiologically, our results imply that E^* can be stable for $\mathcal{R}_0 > 1$ with some realistic parameters, and have broadened the possibilities of application of an age-structured SIR epidemic model for the estimation of $\mathcal{R}_0$ based on the real data of endemic diseases.

In this study, we restricted our attention to the case where the disease transmission function $\kappa = \kappa(a)$ is only dependent on the age a of infective individuals. Of course, the case where it depends on the ages of both susceptible and infective individuals is more general and epidemiologically realistic. Nevertheless, the stability of the endemic equilibrium of age-structured SIR epidemic models has not been clarified well enough even for the former case. In fact, the global asymptotic stability of the endemic equilibrium E^* of such models does not generally hold even if the basic reproduction number $\mathcal{R}_0$ is greater than unity, as a special case where E^* becomes unstable for $\mathcal{R}_0 > 1$ was obtained in [18]. As an important future work, we will seek other forms of $\kappa(\cdot)$ for which we can apply a similar reduction method to ODEs as in this study.

Funding: This research was funded by Grant-in-Aid for Young Scientists (B) of Japan Society for the Promotion of Science (grant number 15K17585).

Conflicts of Interest: The author declares no conflict of interest.

References

1. Bacaër, N. *A Short History of Mathematical Population Dynamics*; Springer-Verlag: London, UK, 2011.
2. Bernoulli, D. Réflexions sur les avantages de l'inoculation. *Mercue de France* **1760**, 173–190.
3. Ross, R. *The Prevention of Malaria*, 2nd ed.; John Murray: London, UK, 1911.
4. Kermack, W.O.; McKendrick, A.G. Contributions to the mathematical theory of epidemics—I. *Proc. R. Soc.* **1927**, *115*, 700–721. [CrossRef]
5. Busenberg, S.N.; Iannelli, M.; Thieme, H.R. Global behavior of an age-structured epidemic model. *SIAM J. Math. Anal.* **1991**, *22*, 1065–1080. [CrossRef]
6. Röst, G.; Wu, J. SEIR epidemiological model with varying infectivity and infinite delay. *Math. Biosci. Eng.* **2008**, *5*, 389–402. [PubMed]
7. Enatsu, Y.; Nakata, Y.; Muroya, Y. Global stability of SIRS epidemic models with a class of nonlinear incidence rates and distributed delays. *Acta Math. Sci.* **2012**, *32*, 851–865. [CrossRef]
8. Iannelli, M. *Mathematical Theory of Age-Structured Population Dynamics*; Giardini Editori e Stampatori: Pisa, France, 1995.
9. Inaba, H. *Age-Structured Population Dynamics in Demography and Epidemiology*; Springer: Singapore, 2017.
10. Ruan, S. Spatial-temporal dynamics in nonlocal epidemiological models. In *Mathematics for Life Science and Medicine*; Takeuchi, Y., Iwasa, Y., Sato, K., Eds.; Springer: Berlin/Heidelberg, Germany, 2007; pp. 97–122.
11. Frasca, M.; Buscarino, A.; Rizzo, A.; Fortuna, L.; Boccaletti, S. Dynamical network model of infective mobile agents. *Phys. Rev. E* **2006**, *74*, 036110. [CrossRef] [PubMed]

12. Diekmann, O.; Heesterbeek, J.A.J.; Metz, J.A.J. On the definition and the computation of the basic reproduction ratio R_0 in models for infectious diseases in heterogeneous populations. *J. Math. Biol.* **1990**, *28*, 365–382. [CrossRef] [PubMed]

13. Hethcote, H.W. Qualitative analyses of communicable disease models. *Math. Biosci.* **1976**, *28*, 335–356. [CrossRef]

14. Gumel, A.B. Causes of backward bifurcations in some epidemiological models. *J. Math. Anal. Appl.* **2012**, *395*, 355–365. [CrossRef]

15. Greenhalgh, D. Hopf bifurcation in epidemic models with a latent period and nonpermanent immunity. *Math. Comput. Model.* **1997**, *25*, 85–107. [CrossRef]

16. Greenhalgh, D. Threshold and stability results for an epidemic model with an age-structured meeting rate. *IMA J. Math. Med. Biol.* **1988**, *5*, 81–100. [CrossRef]

17. Inaba, H. Threshold and stability results for an age-structured epidemic model. *J. Math. Biol.* **1990**, *28*, 411–434. [CrossRef] [PubMed]

18. Thieme, H. Stability change of the endemic equilibrium in age-structured models for the spread of S-I-R type infectious diseases. In *Differential Equations Models in Biology, Epidemiology and Ecology*; Busenberg, S., Martelli, M., Eds.; Springer: Berlin/Heidelberg, Germany, 1991; Volume 92, pp. 139–158.

19. Andreasen, V. Instability in an SIR-model with age-dependent susceptibility. In *Mathematical Population Dynamics: Analysis of Heterogeneity*; Arino, O., Axelrod, D., Kimmel, M., Langlais, M., Eds.; Wuerz Publishing: Winnipeg, MB, Canada, 1995; pp. 3–14.

20. Cha, Y.; Iannelli, M.; Milner, F.A. Stability change of an epidemic model. *Dyn. Syst. Appl.* **2000**, *9*, 361–376.

21. Franceschetti, A.; Pugliese, A.; Breda, D. Multiple endemic states in age-structured SIR epidemic models. *Math. Biosci. Eng.* **2012**, *9*, 577–599. [CrossRef] [PubMed]

22. Ministry of Health, Labour and Welfare, the 22nd Life Tables. Aviliable online: https://www.mhlw.go.jp/english/database/db-hw/lifetb22nd/index.html (accessed on 12 July 2018).

23. Centers for Disease Control and Prevention, Clinical Signs and Symptoms of Influenza. Aviliable online: https://www.cdc.gov/flu/professionals/acip/clinical.htm (accessed on 13 July 2018).

24. Mills, C.E.; Robins, J.M.; Lipsitch, M. Transmissibility of 1918 pandemic influenza. *Nature* **2004**, *432*, 904–906. [CrossRef] [PubMed]

25. Heijne, J.C.M.; Herzog, S.A.; Althaus, C.L.; Low, N.; Kretzschmar, M. Case and partnership reproduction numbers for a curable sexually transmitted infection. *J. Theor. Biol.* **2013**, *331*, 38–47. [CrossRef] [PubMed]

![mathematics logo] *mathematics*

MDPI

Article

Bifurcation Analysis of a Certain Hodgkin-Huxley Model Depending on Multiple Bifurcation Parameters

André H. Erhardt [ID]

Department of Mathematics, University of Oslo, P.O. Box 1053 Blindern, 0316 Oslo, Norway;
andreerh@math.uio.no

Received: 10 May 2018; Accepted: 12 June 2018; Published: 18 June 2018

check for updates

Abstract: In this paper, we study the dynamics of a certain Hodgkin-Huxley model describing the action potential (AP) of a cardiac muscle cell for a better understanding of the occurrence of a special type of cardiac arrhythmia, the so-called early afterdepolarisations (EADs). EADs are pathological voltage oscillations during the repolarisation or plateau phase of cardiac APs. They are considered as potential precursors to cardiac arrhythmia and are often associated with deficiencies in potassium currents or enhancements in the calcium or sodium currents, e.g., induced by ion channel diseases, drugs or stress. Our study is focused on the enhancement in the calcium current to identify regions, where EADs related to enhanced calcium current appear. To this aim, we study the dynamics of the model using bifurcation theory and numerical bifurcation analysis. Furthermore, we investigate the interaction of the potassium and calcium current. It turns out that a suitable increasing of the potassium current adjusted the EADs related to an enhanced calcium current. Thus, one can use our result to balance the EADs in the sense that an enhancement in the potassium currents may compensate the effect of enhanced calcium currents.

Keywords: nonlinear dynamics; bifurcation analysis; ion current interactions; EADs; MATCONT

MSC: 37G15; 37N25; 65P30; 92B05

1. Introduction

The aim of this manuscript is the mathematical and numerical investigation of a three-dimensional Hodgkin-Huxley model from [1] to study early afterdepolarisations (EADs) in a cardiac muscle cell. Here, we are interested in the dynamics of this system and mainly in reasons for the occurrence of EADs. More precisely, we are interested in the sudden change from a normal action potential (AP) to this special type of cardiac arrhythmia. In general, EADs are additional small amplitude spikes during the plateau or the repolarisation phase of the AP. The presence of EADs strongly correlates with the onset of dangerous cardiac arrhythmias, including torsades de pointes (TdP), which is a specific type of abnormal heart rhythm that can lead to sudden cardiac death, see [2–4]. Please see for more (biological/physiological) details [5–11]. In this paper, we will use the bifurcation analysis similar to [12–14] to study the system introduced in [1]. We want to highlight that we can use our approach to investigate also more complex models, see for instance [15–17]. The numerical effort will be higher but we can use this basic principle for further studies. Moreover, our approach can be utilised to study the ion current interaction of all ion currents; this approach is not restricted to the investigation of the potassium–calcium current interaction. In addition, the general aim is to extend this research to models of the complete heart, cf. [18,19]. The main novelty of this paper is the consideration of a bifurcation problem depending on two bifurcation parameters to investigate the ion current interaction and the occurrence of EADs related to the calcium current. Furthermore, we want to mention that the intention of our manuscript is to present our results to a wide range of scientific researcher with different background in biology, mathematics, physics and physiology.

In last decades, the mathematical investigation of phenomena from life science, especially mathematical modelling and mathematical analysis, aroused more and more importance, as well as interdisciplinary research involving mathematics. A very important topic in recent years is the mathematical investigation of (human) diseases, e.g., tumor growth as well as diseases in neurons or cardiac muscle cells. Also here we have the aim to push forward this progress using our knowledge from mathematics to refine and extend existing theory and results. To this goal we choose the model from [1] for our study of EADs, where the authors studies two types of cardiac arrhythmias for instance EADs related to a deficit in the potassium current. Later we will give more details on their approach and how it differs from our ansatz, please see Section 3.2. In this paper, the authors chose the model from [20] and considered a modified reduced version, which they final reduced to a two-dimensional model (fast subsystem) for their investigation. Using their approach the authors in [1] have shown that EADs can occur during the transition between stable steady states. Moreover, they argued their study shows that a stable limit cycle in the fast subsystem is not required for EAD generation. In Section 3.2, we will explain and show why this approach is not applicable to study and to understand EADs related to an enhancement in the calcium current.

Moreover, there are several cardiac cell models available, e.g., the famous Luo-Rudy model [17] describing a cardiac muscle cell of guinea pigs. Furthermore, we want to refer to the review article on cardiac cell modelling [16], where the authors give a nice overview on existing cardiac cell models, as well as to the review article on cardiac tissue modelling [15].

In addition to the modelling of phenomena in the real life it is very important to analyse the behaviour of the corresponding model and to study its dynamics. To this aim we are using the bifurcation theory, since this theory provides a strategy for investigating the bifurcations and the behaviour of the system, please see Section 3. Furthermore, we want to highlight the monographs [12–14]. These monographs give a very good introduction and nice overview on this topic. In [12] the author does not only explain and discuss the bifurcation theory, he also provides the numerical background for the numerical bifurcation analysis. Moreover, the books [13,14] are focused on the qualitative study of high-dimensional nonlinear dynamical systems and chaos. Beside these books there are plenty of further good books dealing with the topics of dynamical systems, bifurcations and chaos, but we cannot cite them all in this manuscript.

The paper is organised as follows. First of all in Section 2, we will give a brief introduction into the topic of cardiac APs and arrhythmia, i.e., afterdepolarisations. Then, we will go on with the mathematical modelling of the cardiac AP using a Hodgkin-Huxley type formalism, which is the usual approach for the modelling of AP for neurons and cardiac muscle cells. In Sections 3 and 4, we will explain the behaviour of the considered model using the bifurcation analysis. The desired bifurcation diagram we will derive utilising MATLAB together with the toolboxes MATCONT and CL_MATCONT [21–23], which are numerical continuation packages for the interactive bifurcation analysis of dynamical systems. In Section 5, we show how we can compensate or control the occurrence of EADs. Finally, we will finish this paper with a discussion.

2. Biological Background and Mathematical Modelling

Action potential. An AP is a temporary, characteristic variance of the membrane potential of an excitable biological cell, e.g., neuron or cardiac muscle cell, from its resting potential. The molecular mechanism of an AP is based on the interaction of voltage-sensitive ion channels. The reason for the formation and the special properties of the AP is established in the properties of different groups of ion channels in the plasma membrane. An initial stimulus activates the ion channels as soon as a certain threshold potential is reached. Then, these ion channels break open and/or up such that this interaction allows an ion current, which changes the membrane potential. A normal AP is always uniform and the cardiac muscle cell AP is typically divided in four phases, i.e., the resting phase, the upstroke phase, the (long) plateau phase and the repolarisation phase, see Figure 1:

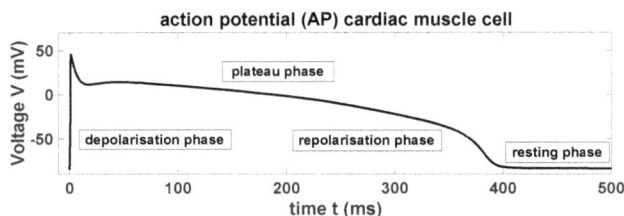

Figure 1. One example of a normal characteristic action potential (AP) of a cardiac muscle cell.

The resting phase/potential is designated by high potassium (K^+) currents, while after the initial stimulus the sodium (Na^+) conductance increases rapidly and the Na^+ current flux into the cardiac muscle cell until a spike potential (ca. +30 mV) is achieved. This spike potential is the so-called upstroke or overshot. Then, the Na^+ current inactivates rapidly followed by the activation of L-type calcium (Ca^{2+}) current. The Ca^{2+} current is more slowly than the Na^+ current and plays a key role in maintaining the long plateau phase, which is characteristic for the cardiac muscle cell. The overall duration of this long AP is at 220–400 ms. Moreover, while the Ca^{2+} conductance increases the K^+ conductance decreases. The plateau phase is followed by the repolarisation phase, where the intrinsic K^+ ion channels are activated and this is connected with the reduction of the Ca^{2+} conductance. Finally, the K^+ current increases until the resting potential respectively the resting phase is reached. In contrast to the Na^+ and Ca^{2+} currents is the K^+ current an outward current.

Afterdepolarisation. If there are depolarising variations of the membrane voltage, then we are speaking about afterdepolarisations. These afterdepolarisations are divided in early afterdepolarisations (EADs) and delayed afterdepolarisations (DADs). This division depends on the timing obtaining of the AP. EADs occur either in the plateau or the repolarisation phase of the AP. EADs are benefited by an elongation of the AP, while the DADs occur after the repolarisation phase is completed. EADs are additional small amplitude spikes during the plateau or the repolarisation phase of the AP, cf. Figure 2.

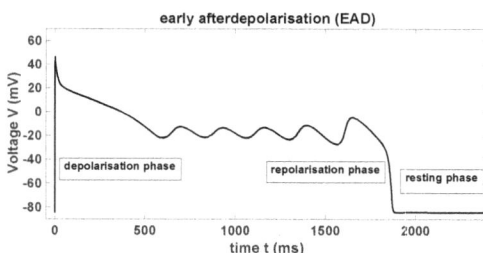

Figure 2. One example of an early afterdepolarisations (EADs).

They are resulting, e.g., from a reduction of the repolarising K^+ currents or from an intensification of depolarising Na^+ currents or Ca^{2+} currents. Triggers for this are congenital disorders of the ion channels (congenital Long- QT-Syndrome) or the ingestion of some medicaments. The elongation of the AP could generate afterpolarisations by reactivation L-type Ca^{2+} influx. Also chronic cardiac insufficiency could appear with an elongation of the AP by a reduction of the repolarising K^+ currents. This means, EADs are often associated with deficits in the potassium currents or enhancements in the calcium currents.

The model. In general, APs of excitable biological cells such as neurons and cardiac muscle cells are often modelled as an ODE system using a Hodgkin-Huxley type formalism, please see the paper of Hodgkin and Huxley [24] and the book of Izhikevich [25]. In this paper, we will study the model from [1]—a toy model—which depends only on two ion currents, the potassium (K$^+$) current and the calcium (Ca^{2+}) current, as a simplification for the study of EADs. Please notice that in this model the sodium current is not considered. Furthermore, the system is self-oscillating, which means that we do not need an initial stimulus. Nevertheless, this is a suitable starting point for the study of EADs, since EADs appear either in the plateau or the repolarisation phase. Here, the potassium (K$^+$) and the calcium (Ca^{2+}) currents are important, see [1,26]. The model is a three-dimensional ODE system, which contains the potassium current $I_K = G_K \cdot x \cdot (V - E_K)$ and the calcium current $I_{Ca} = G_{Ca} \cdot f \cdot d_\infty(V) \cdot (V - E_{Ca})$, where V denotes the membrane voltage, $d_\infty(V)$ is the steady state of the gating variable d given in (2), $G_K = 0.05 \frac{mS}{cm^2}$ and $G_{Ca} = 0.025 \frac{mS}{cm^2}$ denote the ion current conductances, while x and f represent the gating variables, which are important for the opening and closing of the different ion channels. Moreover, E_K and E_{Ca} denote the Nernst potential of the ion currents, cf. Table 1. The physical system (Figure 3) one has in mind is the following

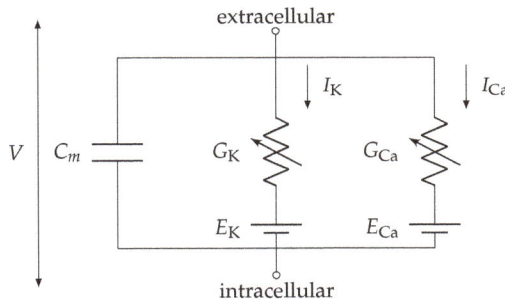

Figure 3. Physical system.

and the corresponding mathematical model read as follows:

$$\begin{cases} \dfrac{dV}{dt} &= -\dfrac{I_K + I_{Ca}}{C_m} =: F_1(V, f, x), \\[2mm] \dfrac{df}{dt} &= \dfrac{f_\infty(V) - f}{\tau_f} =: F_2(V, f, x), \\[2mm] \dfrac{dx}{dt} &= \dfrac{x_\infty(V) - x}{\tau_x} =: F_3(V, f, x), \end{cases} \tag{1}$$

where $\tau_f = 80$ ms and $\tau_x = 300$ ms denote the relaxation time constant of the corresponding channel gating variables. Further, the membrane capacitance is given by $C_m = 1 \, \mu F/cm^2$, cf. [1,20,27]. As we already mentioned, the gating variables are important for the opening and closing of the different ion channels. This means (for instance explained for I_K and $x \in [0, 1]$) that if $x = 0$ the potassium channel is closed, i.e., there is no potassium current flow ($I_K = 0$), while if $x = 1$ the potassium channel is complete open. We have also to mention that the calcium current is depending on a second gating variable d, which is assumed to be equal to its equilibrium, cf. (2).

Conductance-based models are based on an equivalent circuit representation of a cell membrane. These models represent a minimal biophysical interpretation for an excitable biological cell in which current flow across the membrane is due to charging of the membrane capacitance and movement of ions across ion channels. Ion channels are selective for particular ionic species, such as calcium or potassium, giving rise to currents I_{Ca} or I_K, respectively. In addition, the equilibria of the gating variables are represented by

$$y_\infty(V) := \left(1 + \exp\left(\frac{V - V_{T_y}}{k_y}\right)\right)^{-1}, \tag{2}$$

where y represents the gating variables d, f and x we will use the abbreviations $y_\infty := y_\infty(V)$ with

Table 1. System parameters of model (1).

Abbr.	Value	Unit	Abbr.	Value	Unit	Abbr.	Value	Unit
V_{T_d}	−35.0	mV	k_d	−6.24	mV	E_K	−80.0	mV
V_{T_f}	−20.0	mV	k_f	8.6	mV	E_{Ca}	100.0	mV
V_{T_x}	−40.0	mV	k_x	−5	mV	V	(−80; 100)	mV

In this setting there are no EADs, but it is well known that the reduction of the potassium or the enhancements in the calcium current or a combination of both may yield EADs. In this manuscript we are focused on the enhancements in the I_{Ca} current and the resulting behaviour of system (1).

3. Bifurcation Theory

A bifurcation of a dynamical system is a qualitative change in its dynamics produced by varying parameters. We consider an autonomous system of ordinary differential equations, where the right hand side of this system is depending on several state variables and parameter(s), cf. system (1). V, f and x denote the state variables, while G_K and G_{Ca} are the parameters of our interest. Since we study the occurrence of EADs induced by an enhancement in the calcium current I_{Ca}, we will choose the conductance G_{Ca} as the bifurcation parameter to be able to simulate the decreasing or mainly the increasing of the current I_{Ca}. In general, a bifurcation occurs at some parameter p (in our case G_{Ca}), if there are parameter values arbitrarily close to p with dynamics topologically inequivalent from those at p. For example, an equilibrium of the considered system may lose or win stability at a bifurcation, or a limit cycle may occur. The bifurcation theory provides a strategy for investigating the bifurcations and the behaviour of the system. This basic idea we will use to study the dynamics of (1) and the reasons for the appearing of EADs. Therefore, we choose the conductance G_{Ca} as bifurcation parameter and we start determining the equilibrium of model (1). This yields $f \equiv f_\infty(V)$ and $x \equiv x_\infty(V)$. Moreover, we have the following condition for the equilibrium of the voltage V:

$$-G_K \cdot x_\infty(V) \cdot (V - E_K) - G_{Ca} \cdot f_\infty(V) \cdot d_\infty(V) \cdot (V - E_{Ca}) = 0. \tag{3}$$

Please note that in system (1) we have at least four system parameters, which are important for the behaviour of the system, i.e., G_{Ca}, G_K, τ_f and τ_x. Here, we are focused mainly on the dependences on G_{Ca} but also on G_K. Further, we want to emphasise, if we change τ_f and/or τ_x, then this has also an effect on the behaviour of the model (1).

At this stage, we see—cf. condition (3)—that the choice of G_{Ca} has a direct influence on the location of the equilibrium and its stability (considering the corresponding eigenvalues). Therefore, varying the bifurcation parameter G_{Ca} yields different equilibria with probably different stability. Figure 4 shows the bifurcation diagram of system (1), where we use G_{Ca} as the bifurcation parameter for fixed $G_K = 0.05 \frac{mS}{cm^2}$. For our bifurcation analysis and the calculation of our bifurcation diagram we are using MATCONT in combination with MATLAB. Please notice that all figures are produced using MATLAB. Figure 4a shows the bifurcation diagram (G_{Ca}, V) with two limit cycle branches, while Figure 4b shows the bifurcation diagram (G_{Ca}, f, V) with (only) the first limit cycle branch. Furthermore, in Figure 4c we state the zoom of Figure 4a around the LPCs (lower limit cycle branches) including two vertical dashed lines for an easier comprehension.

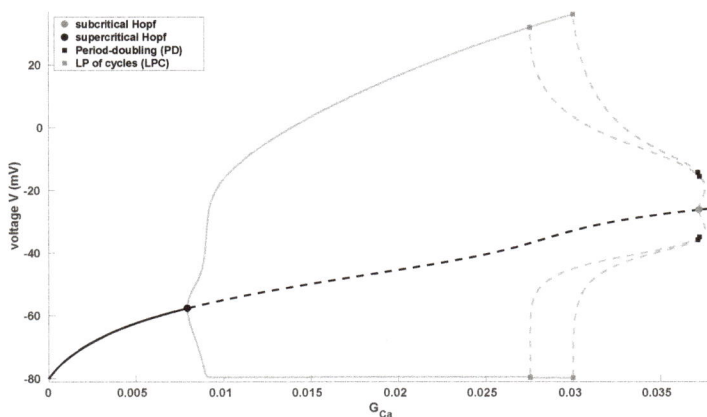

(**a**)2D bifurcation diagram.

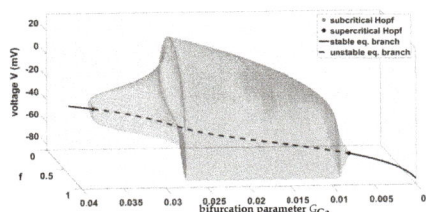

(**b**)3D bifurcation diagram.

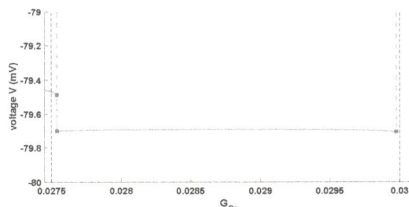

(**c**)Zoom of Figure 4a around the LPCs (lower branch).

Figure 4. Bifurcation diagram of system (1) with G_{Ca} as bifurcation parameter. The black line denotes the stable branch of the equilibrium curve, while the black dashed line the unstable one. The grey line and grey dashed line represent the stable or unstable limit cycle branches, respectively.

This bifurcation diagram shows that the equilibrium curve loses stability via a subcritical Andronov-Hopf bifurcation (grey dot) and wins stability via a supercritical Andronov-Hopf bifurcation (black dot)—Figure 4a from right to left and Figure 4b from left to right. Moreover, an unstable limit cycle branch bifurcates from the subcritical Andronov-Hopf bifurcation (positive first Lyapunov coefficient), which becomes stable via a limit point bifurcation (LP) of cycles (also known as fold or saddle-node bifurcation of cycles, which generically corresponds to a turning point of a curve of limit cycles) and finally, disappears via the supercritical Andronov-Hopf bifurcation (negative first Lyapunov coefficient). Please notice that after the supercritical Andronov-Hopf bifurcation the steady state becomes an unstable saddle-focus for G_{Ca} values approximately between (0.080; 0.0138). For values approximately between (0.0138; 0.0320) it turns into a saddle before it becomes again an unstable saddle-focus and gains again stability via the subcritical Andronov-Hopf bifurcation. Please note that the saddle has always two positive and one negative (real) eigenvalue, i.e., we have a two-dimensional unstable and an one-dimensional stable manifold. From this bifurcation diagram it is obvious that oscillations (in the sense of periodic orbits, which are not converging into an equilibrium or which no more reach the resting potential) can occur only between the two Andronov-Hopf bifurcations—also trajectories, which are not related to EADs. Notice that the system (1) exhibits also oscillations for values of G_{Ca} close to the right hand side of the subcritical Andronov-Hopf bifurcation, but the trajectory will either converge into a stable focus after a certain amount of time, which would be related to the sudden death if this behaviour spread over the heart, or oscillations occur, which exhibit no resting phase and oscillate continuously (depending on the initial values). Since the aim of this paper is to identify the region, where EADs appear, to be able to control them and to prevent the sudden death, we are focused on the distinction of the spiking regions (related to normal AP) and the bursting region (with periodic orbits related to EADs). However, EADs are in general complex oscillatory phenomena and could have one or more additional small oscillations. For simplicity we will analyse and point out, where system (1) have no additional small oscillations and for which values of G_{Ca} EADs appear.

At this stage, we have to notice that increasing of G_{Ca} and therefore, increasing of the calcium current I_{Ca} may yields EADs, which was expected. Moreover, this bifurcation diagram implies that EADs can appear only for values of G_{Ca} between the subcritical Andronov-Hopf bifurcation ($G_{Ca} \approx 0.0372 \frac{mS}{cm^2}$) and LP of cycles ($G_{Ca} \approx 0.0275 \frac{mS}{cm^2}$), which is the separatrix between EADs and no EADs, since "after" the first LP of cycles there are no additional small oscillations. From this LP of cycles we have the stable limit cycle branch, while from the first period doubling bifurcation again an unstable limit cycle branch bifurcates, which becomes stable after a further LP of cycles, again unstable via a third LP of cycles before this branch converges into the first unstable limit cycle branch. Please notice that we do not have isolas as in [28] but we also have a similar splitting into a spiking and a bursting region, see Figure 4c and its refined view in Figure 5.

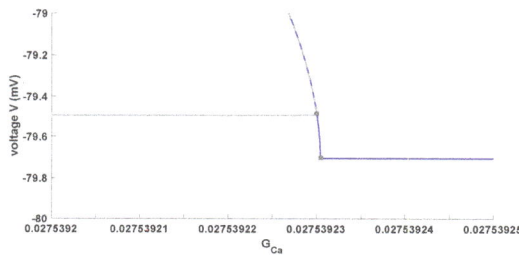

Figure 5. Refined view on Figure 4c at the transition of the first two limit cycle branches.

In Figure 5 we plot the continuation from the first PD (the second limit cycle branch) as a solid blue line to illustrate in a better way the situation we mentioned above. Furthermore, we want to highlight that such models (which exhibit one or more PD) might exhibit isolas as in [28] or a PD cascade as in [27]. In the case that there exists a (stable) PD cascade then the model contains chaos depending on the value of the bifurcation parameter. This is depending on the choices of the system parameters, e.g., τ_f and τ_x. However, the (unforced) system (1) exhibits no chaotic pattern or trajectories. But, chaotic trajectories may occur for different values of τ_f and τ_x. Please see for more details [27], where we have shown that system (1) may have a (stable) PD cascade, which is usually the route to chaos.

3.1. Bifurcation Analysis with G_K as Bifurcation Parameter

Our next observation is, if we choose $G_{Ca} = 0.03 \frac{mS}{cm^2}$ and $G_K = 0.05 \frac{mS}{cm^2}$, since we know from Figure 4c that we have two small oscillations (second LPC at $G_{Ca} \approx 0.029976 \frac{mS}{cm^2}$), i.e, an EAD—cf. Figure 6a—and we use now G_K as bifurcation parameter with fixed $G_{Ca} = 0.03 \frac{mS}{cm^2}$, it turns out that increasing of the potassium current can balance the effect of an enhanced calcium current. From the bifurcation diagram in Figure 6b, we get that there are no EADs for $G_{Ca} = 0.03 \frac{mS}{cm^2}$ and G_K values greater than $G_K \approx 0.0548 \frac{mS}{cm^2}$ (LP of cycles), while for values between $G_K \approx 0.0408 \frac{mS}{cm^2}$ (subcritical Andronov-Hopf bifurcation) and $G_K \approx 0.0548 \frac{mS}{cm^2}$ there are EADs, similar to the discussion from above. Therefore, the effect of an enhanced calcium current I_{Ca} can be compensated by an increasing of the potassium current I_K.

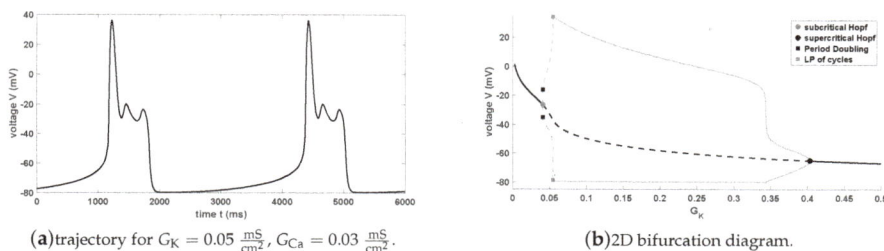

(**a**) trajectory for $G_K = 0.05 \frac{mS}{cm^2}$, $G_{Ca} = 0.03 \frac{mS}{cm^2}$.

(**b**) 2D bifurcation diagram.

Figure 6. Bifurcation diagram of (1) for the case $G_{Ca} = 0.03 \frac{mS}{cm^2}$ and G_K used as bifurcation parameter.

Regarding Figure 6b, we see that if we choose a G_K value too close to the supercritical Andronov-Hopf bifurcation, then the voltage does not reach the resting potential. This indicates that we have normal AP as long as we choose a G_K value such that this value is greater than the value of the LP of cycles and the lower branch of the limit cycle branch is equal to the resting potential of the voltage.

3.2. Multiple Time Scales

Here, we want to remark that for instance in [1,26] the occurrence of EADs via the reduction of the potassium current is studied. The authors used a time scale separation argument (not explicit) to identify the gating variable x as the slowest variable and then, they argued in principle that EADs are Hopf-induced, cf. [29–31], by considering a fast subsystem using x as bifurcation parameter. This approach is not applicable in our situation, since the gating variable f is much faster than x. This one can realise since $\tau_f \ll \tau_x$, but we can show this also by the following time scale separation argument. To this aim we introduce a new (dimensionless) time variable τ satisfying $t := k_t \cdot \tau$, where k_t is a reference time we have to choose. Choosing $k_t = \tau_x$ and re-writing system (1) we get:

$$
\begin{cases}
\varepsilon \dfrac{dV}{d\tau} &= -(I_K + I_{Ca}), \\[2mm]
\delta \dfrac{df}{d\tau} &= (f_\infty(V) - f), \\[2mm]
\dfrac{dx}{d\tau} &= (x_\infty(V) - x),
\end{cases}
\tag{4}
$$

where we divided also the first equation by $G := \max\{G_K, G_{Ca}\}$ and defined $\bar{G}_K := \frac{G_K}{G}$ and $\bar{G}_{Ca} := \frac{G_{Ca}}{G}$ to derive the dimensionless singular perturbation parameters $\varepsilon := \frac{C_m}{\tau_x \cdot G}$ and $\delta := \frac{\tau_f}{\tau_x}$. Please note that we chose as reference time k_t the maximum of all relaxation time constants. Using the setting from above we have that $0 \leq \varepsilon < \delta \ll 1$, i.e., three different time scales. Furthermore, we want to highlight that also different choices of the singular perturbation parameters yield that the gating variable x is always the slowest variable and the corresponding fast subsystem is either 1 dimensional (if $\varepsilon \to 0$ and $\delta \neq 0$ fixed), i.e., it cannot exhibit an Andronov-Hopf bifurcation, or 2 dimensional (if $\delta \equiv \varepsilon$, $\varepsilon \to 0$):

$$
\begin{cases}
\dfrac{dV}{d\tau_{\text{fast}}} &= -(I_{\text{K}} + I_{\text{Ca}}), \\[2ex]
\dfrac{df}{d\tau_{\text{fast}}} &= \dfrac{G}{\tau_f C_m}(f_\infty(V) - f), \\[2ex]
x &= \text{const.},
\end{cases}
$$

where we rescaled in time, i.e., $\tau_{\text{fast}} = \tau/\varepsilon$, and considered the singular limit $\tau_x \to \infty$ yielding $\varepsilon \to 0$. Therefore, EADs appearing in system (1) via an enhanced calcium current are not Hopf-induced in the sense of geometric singular perturbation theory, please see the book of Kuehn [31] for more details. Again, we want to highlight that EADs may occur as Hopf-induced mixed mode oscillations via a reduction of the potassium current, but not via an enhancement in the calcium current.

4. Two bifurcation problem

Our next aim is to use the previous result to investigate the ion current interactions by considering a two bifurcation problems. First of all, we have to notice that varying simultaneous the conductances G_{K} and G_{Ca} yields the equilibrium curves in Figure 7. For suitable values of G_{K} and G_{Ca} the stable equilibrium branch loses stability via an Andronov-Hopf bifurcation (sub- or supercritical) and becomes stable via a further (supercritical) Andronov-Hopf bifurcation. The range, where system (1) oscillates, increases for increased values of G_{K} and G_{Ca}.

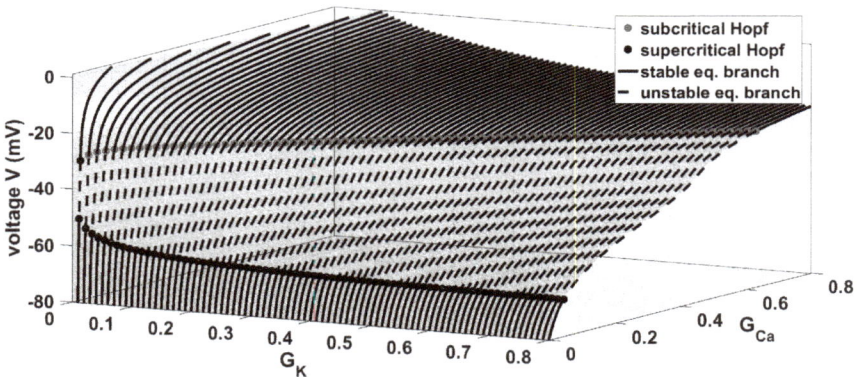

Figure 7. Bifurcation problem with G_{K} and G_{Ca} as bifurcation parameters.

This has also a huge influence on the behaviour of the complete system. Moreover, for each new limit cycle branch the trajectory has a further small oscillation (via a PD cascade or isolas), depending on the choice of G_{Ca}. This means—cf. Figure 4—that the trajectory has one small oscillation for G_{Ca} values approximately between [0.02754; 0.02997]. At this stage, we emphasise that our approach can be extended to a multiple bifurcation problem of system (1) with up to five main important parameters, i.e., G_{K}, G_{Ca}, C_m, τ_f and τ_x. While the shape of the equilibrium curves are depending on G_{K} and G_{Ca}, the shape of the equilibrium curves remain the same by varying C_m, τ_f or τ_x, but the behaviour is depending on these parameters, cf. Figure 8. Moreover, the Andronov-Hopf bifurcations in Figures 7 and 8 form two "Hopf-curves", which continuously depend on G_K and G_{Ca}. This we can also prove by the Routh-Hurwitz criterion, see [13]. For this aim we determine the characteristic equation

$$-\det(A - \lambda \mathbb{1}_3) = \lambda^3 + a_1 \lambda^2 + a_2 \lambda + a_3 = 0, \tag{5}$$

where A denotes the Jacobian of system (1) evaluated at the equilibrium of system (1). The Routh-Hurwitz criterion implies that all characteristic exponents of Equation (5) have negative real parts if and only if the conditions

$$\Delta_1 = a_1 > 0, \quad \Delta_2 = a_1 a_2 - a_3 > 0 \quad \text{and} \quad \Delta_3 = a_3(a_1 a_2 - a_3) > 0$$

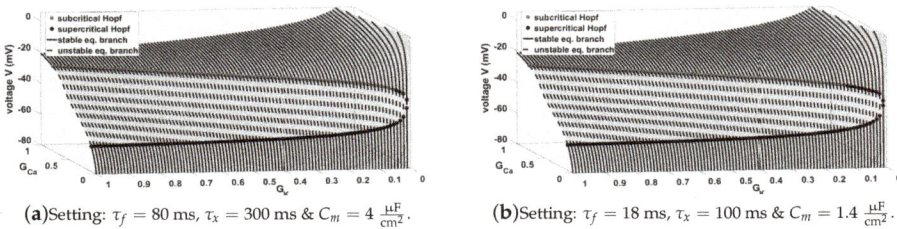

(a) Setting: $\tau_f = 80$ ms, $\tau_x = 300$ ms & $C_m = 4 \frac{\mu F}{cm^2}$.

(b) Setting: $\tau_f = 18$ ms, $\tau_x = 100$ ms & $C_m = 1.4 \frac{\mu F}{cm^2}$.

Figure 8. 2-bifurcation problem: equilibrium curves of (1) for different settings showing the shape of the equilibrium curves remain the same but the behaviour of the curves is depending on the choices of the parameters τ_f, τ_x and C_m.

are satisfied, which implies that the equilibrium is asymptotically stable, where $a_1 = \left(\frac{1}{\tau_f} + \frac{1}{\tau_x} - \frac{\partial F_1}{\partial V}\right)$,
$a_2 = \left(\frac{1}{\tau_f \tau_x} - \left(\frac{1}{\tau_f} + \frac{1}{\tau_x}\right)\frac{\partial F_1}{\partial V} - \frac{1}{\tau_f}\frac{\partial f_\infty}{\partial V}\frac{\partial F_1}{\partial f} - \frac{1}{\tau_x}\frac{\partial x_\infty}{\partial V}\frac{\partial F_1}{\partial x}\right)$, $a_3 = -\frac{1}{\tau_f \tau_x}\left(\frac{\partial x_\infty}{\partial V}\frac{\partial F_1}{\partial x} + \frac{\partial f_\infty}{\partial V}\frac{\partial F_1}{\partial f} + \frac{\partial F_1}{\partial V}\right)$ and

$$\Delta_2 = \frac{1}{\tau_f}\left(\frac{1}{\tau_f} - \frac{\partial F_1}{\partial V}\right)\left(\frac{1}{\tau_x} - \frac{\partial F_1}{\partial V} - \frac{\partial f_\infty}{\partial V}\frac{\partial F_1}{\partial f}\right) + \frac{1}{\tau_x}\left(\frac{1}{\tau_x} - \frac{\partial F_1}{\partial V}\right)\left(\frac{1}{\tau_f} - \frac{\partial F_1}{\partial V} - \frac{\partial x_\infty}{\partial V}\frac{\partial F_1}{\partial x}\right).$$

Moreover, if $\Delta_1 > 0$, $\Delta_2 = 0$ and $a_3 > 0$ the equilibrium of system (1) is an Andronov-Hopf bifurcation with $\lambda_{1,2} = \pm i\omega_0$, where $\omega_0^2 = a_2 > 0$ and $\lambda_3 = -a_1$, since

$$\lambda^3 + a_1 \lambda^2 + a_2 \lambda + a_3 = \lambda_3 + a_1 \lambda^2 + a_2 \lambda + a_1 a_2 = (\lambda^2 + a_2)(\lambda + a_1) = 0,$$

where we used $\Delta_2 = 0$, cf. [14]. Furthermore, $a_1 > 0$ implies that λ_3 is negative. Especially, if $\Delta_2 < 0$ oscillations occur, which implies that $X(t) - X(t + T) = 0$, where $X = \{V, f, x\}$ and T denotes the period of the periodic orbit. Thus, the conditions $a_1 > 0$, $\Delta_2 = 0$ and $a_3 > 0$ with condition (3), $f = f_\infty$ and $x = x_\infty$ yields (numerically) in dependence on G_K and G_{Ca} the "Hopf-curves" in Figure 7 or Figure 9, respectively. Mainly, we are interested in the separatrix between EADs and no EADs.

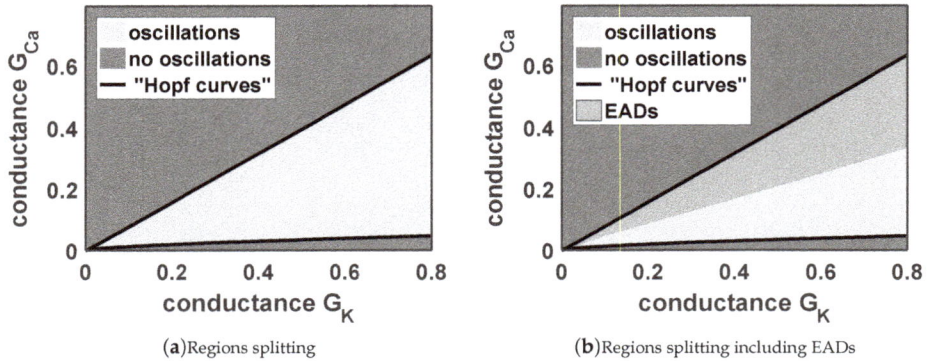

(**a**)Regions splitting

(**b**)Regions splitting including EADs

Figure 9. The two black lines represent the two "Hopf-curves". Between these lines oscillations occur (light grey area), while outside this area (dark grey areas) no oscillations (in the sense we mentioned in Section 3) appear. Here, we have stable equilibria of system (1). Furthermore, the third area in Figure 9b represents the values of G_K and G_{Ca}, where EADs appear.

In Figure 9 we see that also the region, where EADs appear, is linearly depending on G_{Ca} and G_K. Moreover, the dangerous region is also growing if we increase the two conductances G_{Ca} and G_K. Nevertheless, this investigation shows that one can balance EADs utilising the ion current interaction.

5. Controlling the Early Afterdepolarisations

Our final aim is to control the EADs. To this goal we use our observations from above and the knowledge of the ion current interaction. From Figure 9 we know how we have to shift G_K to smooth out the impact of the enhanced calcium current. However, there are several possibilities to achieve this. First, we can control the EADs by varying the conductance G_K in the potassium current I_K (in the case that the choice of G_{Ca} yields an EAD) or we can introduce a control parameter $p_{control}$ and replace G_K by

$$\bar{G}_K := (G_K - p_{control}). \tag{6}$$

Then, using $p_{control}$ as control or bifurcation parameter, respectively, we can balance the EAD. A further approach is to replace G_{Ca} and G_K by

$$\bar{G}_{Ca} := (G_{Ca} + p_{control}) \quad \text{and} \quad \bar{G}_K := (G_K - p_{control}), \tag{7}$$

respectively. In this case we control simultaneous both ion currents I_{Ca} and I_K, i.e., we exploit the ion current interaction. Here, we are now able to study the two bifurcation parameter problem from above, but in this bifurcation problem we only vary one bifurcation parameter, i.e., the control parameter. Moreover, we want to emphasise that we shift G_{Ca} and G_K simultaneous. This yields the same, which we showed exemplary in Figure 6 and corresponds to the regions in Figure 9. Furthermore, we have to highlight that we have to choose the sign of $p_{control}$ opposed in $\bar{G}_{Ca}$ and $\bar{G}_K$, since we know that decreasing of I_K and increasing of I_{Ca} may yield in both cases EADs. This means that these ion currents behave in same sense reverse (outward and inward currents) and this we want to exploit. The advantage in this approach is also that we do not have to change one current "dramatically". Please notice in both cases using (6) or (7) we get different values of $p_{control}$. However, utilising the definitions of $\bar{G}_K$ (and $\bar{G}_{Ca}$) in (6) or (7) and relabelling $\bar{G}_K$ (and $\bar{G}_{Ca}$) by G_K (and G_{Ca})

we get the conclusion of Figure 9. Finally, we are utilising a further approach, i.e., we consider the following system

$$
\begin{cases}
\dfrac{dV}{dt} = -\dfrac{\bar{I}_K + I_{Ca}}{C_m} =: F_1(V, f, x), \\[2mm]
\dfrac{df}{dt} = \dfrac{f_\infty(V) - f}{\tau_f} =: F_2(V, f, x), \\[2mm]
\dfrac{dx}{dt} = \dfrac{x_\infty(V) - x}{\bar{\tau}_x} =: F_3(V, f, x),
\end{cases}
\tag{8}
$$

with

$$
\bar{I}_K := \bar{G}_K \cdot x \cdot (V - E_K), \quad \bar{G}_K := (G_K - p_{control}), \quad \bar{\tau}_x := 0.75 \cdot (\tau_x + p_{control}).
$$

Here, we again consider only the potassium current to control the effect of the enhanced calcium current. However, we are not only focused on the conductance G_K, we are paying attention also to the relaxation time constant τ_x. This has influence on the gating variable x and therefore, again on the potassium current, i.e., decreasing of $p_{control}$ increases $\bar{G}_K$ and x. The choices of the signs of the control parameter in $\bar{G}_K$ and $\bar{\tau}_x$ is again related to the aim to increase the potassium current. This yields the regions in Figure 10.

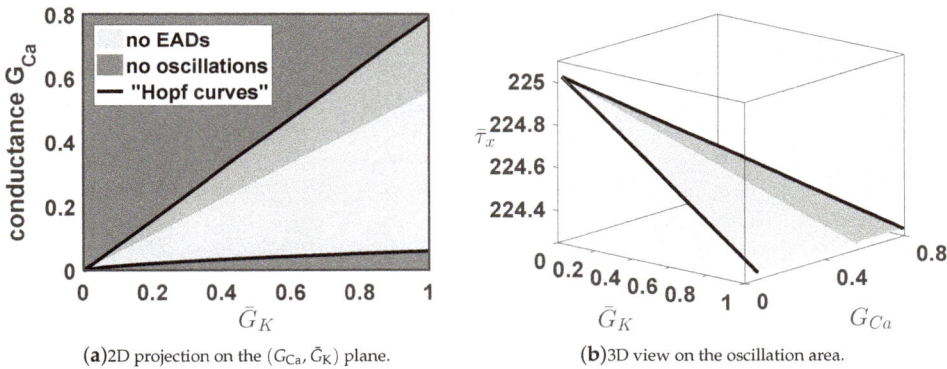

(a) 2D projection on the $(G_{Ca}, \bar{G}_K)$ plane.   **(b)** 3D view on the oscillation area.

Figure 10. The black lines represent the two "Hopf-curves" of system (8) including the oscillatory area.

Here, we see that choosing the approach from (8) yields different region, where oscillations appear, i.e., EADs and no EADs, please cf. Figure 9. We see that – regarding Figure 10a—we have a much bigger range with respect to the conductances G_{Ca} and $\bar{G}_K$, where no EADs appear. To achieve this a reduction of the time relaxation constant τ_x or $\bar{\tau}_x$, respectively, is necessary, cf. Figure 10b. Moreover, as a general remark we want to emphasise that all these approaches can be modified, i.e., we can add weight to the control parameter depending on the system specific properties, as we already did in system (8). In system (8), e.g., we can modify $\bar{G}_K$ and $\bar{\tau}_x$ as follows

$$
\bar{G}_K := (G_K - a \cdot p_{control}) \quad \text{and} \quad \bar{\tau}_x := b \cdot (\tau_x + c \cdot p_{control}),
\tag{9}
$$

where $a, c \in \mathbb{R}^+$, $a \neq c$ and $0 < b$ or similar modifications. Especially, in this paper we are focused on balancing the enhanced calcium current via the increasing the potassium current. Please remark also that all these approaches we have for $p_{control} = 0$ and $b = 1$ the initial situation. Moreover, we know that we can compensate the enhancement of the calcium current by increasing the potassium current.

Furthermore, notice that to balance an EAD induced by an enhanced calcium current the control parameter $p_{control}$ in system (8) has to be negative. Using the approach (8) together with (9) and the choices $a = b = c = 1$ yields also a different regions as in Figure 9, but the "Hopf-curves" are still less steep as in Figure 10. This means we can control the slope of the "Hopf-curves".

Finally, we want to remark that this approach we can also use for more general systems, e.g., including the sodium current I_{Na} or relaxation time constants, which are depending on the voltage V, cf. [17]. This yields then more possible choices of the parameters and the parameter space will grow, but this has also potential to study and to control cardiac arrhythmia in a more specific way.

6. Discussion

In this paper, we studied the Hodgkin-Huxley model from [1] to investigate the occurrence of EADs related to an enhanced calcium current and the ion current interaction of the potassium and calcium current. For this aim we used the bifurcation theory and the numerical bifurcation analysis to derive a separatix between EADs, i.e., mixed mode oscillations [30,31] with one large and one or more small oscillations, and no EADs in system (1), cf. Figure 9. It turns out that our system loses stability and oscillates (in the sense of periodic orbits mentioned in Section 3) between two Andronov-Hopf bifurcations, where EADs as well as no EADs appear (periodic orbits). In this region stable periodic orbits occur, where the number of small oscillations are depending on the numbers of (unstable) limit cycle branches, cf. Figure 4 and [28], where we adumbrated this fact.

Moreover, we showed that no EADs occur in a region between the supercritical Andronov-Hopf bifurcation (related to small values of the conductance G_{Ca}) and the first LP of cycles of the first Hopf continuation—which generically corresponds to a turning point of a curve of limit cycles—from this supercritical Andronov-Hopf bifurcation, cf. Figure 4. This corresponds also with the circumstance that EADs may appear by the enhancement in the calcium current. Please notice that also the phenomena of isolas may occur for different settings of the system parameters, cf. for instance [28], but not in our setting, please cf. Figure 5. Furthermore, we restricted our bifurcation diagram to the first two limit cycle branches. The continuation from an Andronov-Hopf bifurcation is comparably easy and fast, while already the continuation of the second limit cycle branch becomes challenging and needs a lot of computational power. Since we are interested in the region, where EADs appear, and because of these numerical efforts, it makes sense to focus on the first limit cycle branch (mainly if one consider a multiple bifurcation problem).

Furthermore, we highlighted that the effect of an enhanced calcium current can be compensated by an increasing of the potassium current, cf. Figure 6. Since we considered only the toy model (1) depending on two ion currents our study is limited to the ion current interaction of the potassium and the calcium current. However, a similar result, we can expect also from the interaction of the potassium and sodium current. Then, these observations motivate the study of system (1) as a multiple bifurcation problem, i.e., our investigation was not only focused on one bifurcation parameter. Please note that EADs can be induced by an increase of the L-type calcium conductance and by the application of potassium current blockers, cf. [3]. Using two bifurcation parameters, yields that the area, where oscillations in system (1) appear, increases by increasing of G_{Ca} and G_K (simultaneously), see Figure 9.

At this stage we want to emphasise, that the study of model (1) as a multiple bifurcation problem with up to five (most important) parameters yields further cognitions, cf. Figures 8 and 9. Moreover, we can use this approach to study more general Hodgkin-Huxley models or we can utilise this strategy to show that EADs related to a reduction of the potassium current can be compensated by a decreasing of the calcium current.

Summarising, we have shown four main result. Obviously, the increasing of the calcium current may yield EADs. Second, one can use an increased potassium current (e.g., induced by some drug) to compensate EADs derived by an enhanced calcium current (e.g., induced by ion channel disease). On the other hand, one can also balance EADs induced by a reduced potassium current, via the reduction of the calcium current. This means if one is not able to decline the influence of a potassium

Mathematics **2018**, *6*, 103

blocker, one may vanish the effect of the potassium blocker by a "calcium blocker". Third, from our study one can expect that EADs may occur via a combination of an enhanced calcium current and a reduced potassium current, cf. [3]. Therefore, the effect of both phenomena can be mutually reinforcing. Thus, EADs may also appear in a "safe region" focused only on one ion current. Fourth, we showed that EADs related to an enhancement in the calcium current are not Hopf-induced in the sense of the geometric singular perturbation theory [32].

Furthermore, in the paper we used several control approach to balance the effect of the enhancement in the calcium current. Here, it turns out that EADs can be compensated using the approach in (8) in a very effective way. The reason is that we increase simultaneous G_K and τ_x, where we added a weight to τ_x, i.e., we increased the speed of the gating variable x. This yields very different regions if we compare Figures 9b and 10a.

In addition, if we study models containing also a sodium current, then the effects of the different ion currents can be balanced by increasing or decreasing the other ion currents. The investigation of the three main ion currents by considering three parameters G_{Ca}, G_K and G_{Na} will yield a three dimensional parameter space and therefore, we will have more possible choices to prevent the EADs. Even more, we are able to investigate and to understand the ion current interactions by means of the bifurcation theory in a very general way, which is useful for the treatment of cardiac diseases. This short study emphasise the importance of the inclusion of the two or three main ion currents, which are related to the occurrence of EADs and the beneficing of bifurcation analysis in the investigation of cardiac arrhythmia.

Finally, we want to outline that our study was only focused on a toy model of dimension three containing two ion currents and it is as a future project mandatory to consider more complex models to be able to understand all effects yielding EADs. Nevertheless, we can use our approach also in more complex situations for the investigation of EADs. Beside the extension of this study to more general ion current models, one has to think about to the study of multi-domain models to be able to study the effect of EADs on the complete heart. A first good starting point would be to observe a mono-domain model as in [4].

Acknowledgments: The author wishes to thank the anonymous referees for their careful reading of the original manuscript and their comments that eventually led to an improved presentation.

Conflicts of Interest: The author declares no conflict of interest.

References

1. Xie, Y.; Izu, L.T.; Bers, D.M.; Sato, D. Arrhythmogenic Transient Dynamics in Cardiac Myocytes. *Biophys. J.* **2014**, *106*, 1391–1397. [CrossRef] [PubMed]
2. Roden, D.M.; Viswanathan, P.C. Genetics of acquired long QT syndrome. *J. Clin. Investig.* **2005**, *115*, 2025–2032. [CrossRef] [PubMed]
3. Vandersickel, N.; Kazbanov, I.V.; Nuitermans, A.; Weise, L.D.; Pandit, R.; Panfilov, A.V. A Study of Early Afterdepolarizations in a Model for Human Ventricular Tissue. *PLoS ONE* **2015**, *9*. [CrossRef] [PubMed]
4. Vandersickel, N.; Panfilov, A.V. A study of early afterdepolarizations in human ventricular tissue. In Proceedings of the Computing in Cardiology Conference (CinC), Nice, France, 6–9 September 2015; pp. 1213–1216.
5. Bergfeldt, L.; Lundahl, G.; Bergqvist, G.; Vahedi, F.; Gransberg, L. Ventricular repolarization duration and dispersion adaptation after atropine induced rapid heart rate increase in healthy adults. *J. Electrocardiol.* **2017**, *50*, 424–432. [CrossRef] [PubMed]
6. Bondarenko, V.E.; Shilnikov, A.L. Bursting dynamics in the normal and failing hearts. *Sci. Rep.* **2017**, *7*. [CrossRef] [PubMed]
7. Dobrev, D.; Wehrens, X.H.T. Calcium mediated cellular triggered activity in atrial fibrillation. *J. Physiol.* **2017**, *595*, 4001–4008. [CrossRef] [PubMed]
8. Landstrom, A.P.; Dobrev, D.; Wehrens, X.H.T. Calcium Signaling and Cardiac Arrhythmias. *Circ. Res.* **2017**, *120*, 1969–1993. [CrossRef] [PubMed]

9. Sato, D.; Clancy, C.E.; Bers, D.M. Dynamics of sodium current mediated early afterdepolarizations. *J. Clin. Investig.* **2017**, *3*, e00388. [CrossRef] [PubMed]

10. Tsaneva-Atanasova, K.; Shuttleworth, T.J.; Yule, D.I.; Thompson, J.L.; Sneyd, J. Calcium Oscillations and Membrane Transport: The Importance of Two Time Scales. *Multiscale Model. Simul.* **2005**, *3*, 245–264. [CrossRef]

11. Nieuwenhuyse, E.V.; Seemann, G.; Panfilov, A.V.; Vandersickel, N. Effects of early afterdepolarizations on excitation patterns in an accurate model of the human ventricles. *PLoS ONE* **2017**, *12*, e0188867. [CrossRef] [PubMed]

12. Kuznetsov, Y.A. *Elements of Applied Bifurcation Theory*, 2nd ed.; Springer: New York, NY, USA, 1998.

13. Shilnikov, L.P.; Shilnikov, A.L.; Turaev, D.V.; Chua, L.O. *Methods of Qualitative Theory in Nonlinear Dynamics. Part I*; World Scientific: Singapore, 1998; Volume 4.

14. Shilnikov, L.P.; Shilnikov, A.L.; Turaev, D.V.; Chua, L.O. *Methods of Qualitative Theory in Nonlinear Dynamics. Part II*; World Scientific: Singapore, 2001; Volume 5.

15. Clayton, R.H.; Bernus, O.; Cherry, E.M.; Dierckx, H.; Fenton, F.H.; Mirabella, L.; Panfilov, A.V.; Sachse, F.B.; Seemann, G.; Zhang, H. Models of cardiac tissue electrophysiology: Progress, challenges and open questions. *Prog. Biophys. Mol. Biol.* **2011**, *104*, 22–48. [CrossRef] [PubMed]

16. Fink, M.; Niederer, S.A.; Cherry, E.M.; Fenton, F.H.; Koivumäki, J.T.; Seemann, G.; Thul, R.; Zhang, H.; Sachse, F.B.; Beard, D.; et al. Cardiac cell modelling: Observations from the heart of the cardiac physiome project. *Prog. Biophys. Mol. Biol.* **2011**, *104*, 2–21. [CrossRef] [PubMed]

17. Luo, C.H.; Rudy, Y. A model of the ventricular cardiac action potential. Depolarization, repolarization, and their interaction. *Circ. Res.* **1991**, *68*, 1501–1526. [CrossRef] [PubMed]

18. Zimik, S.; Vandersickel, N.; Nayak, A.R.; Panfilov, A.V.; Pandit, R. A Comparative Study of Early Afterdepolarization-Mediated Fibrillation in Two Mathematical Models for Human Ventricular Cells. *PLoS ONE* **2015**, *10*. [CrossRef] [PubMed]

19. Tveito, A.; Jæger, K.H.; Kuchta, M.; Mardal, K.A.; Rognes, M.E. A Cell-Based Framework for Numerical Modeling of Electrical Conduction in Cardiac Tissue. *Front. Phys.* **2017**, *5*. [CrossRef]

20. Sato, D.; Xie, L.-H.; Nguyen, T.P.; Weiss, J.N.; Qu, Z. Irregularly Appearing Early Afterdepolarizations in Cardiac Myocytes: Random Fluctuations or Dynamical Chaos? *Biophys. J.* **2010**, *99*, 765–773. [CrossRef] [PubMed]

21. Dhooge, A.; Govaerts, W.; Kuznetsov, Y.A. MATCONT: A MATLAB Package for Numerical Bifurcation Analysis of ODEs. *ACM Trans. Math. Softw.* **2003**, *29*, 141–164. [CrossRef]

22. Dhooge, A.; Govaerts, W.; Kuznetsov, Y.A.; Meijer, H.G.E.; Sautois, B. New features of the software MatCont for bifurcation analysis of dynamical systems. *Math. Comput. Model. Dyn. Syst.* **2008**, *14*, 147–175. [CrossRef]

23. Govaerts, W.; Kuznetsov, Y.A.; Dhooge, A. Numerical Continuation of Bifurcations of Limit Cycles in MATLAB. *SIAM J. Sci. Comput.* **2005**, *27*, 231–252. [CrossRef]

24. Hodgkin, A.L.; Huxley, A.F. A quantitative description of membrane current and its application to conduction and excitation in nerve. *J. Physiol.* **1952**, *117*, 500–544. [CrossRef] [PubMed]

25. Izhikevich, E.M. *Dynamical Systems in Neuroscience*; The MIT Press: Cambridge, MA, USA, 2007.

26. Tran, D.X.; Sato, D.; Yochelis, A.; Weiss, J.N.; Garfinkel, A.; Qu, Z. Bifurcation and Chaos in a Model of Cardiac Early Afterdepolarizations. *Phys. Rev. Lett.* **2009**, *102*, 258103. [CrossRef] [PubMed]

27. Kügler, P.; Bulelzai, M.A.K.; Erhardt, A.H. Period doubling cascades of limit cycles in cardiac action potential models as precursors to chaotic early Afterdepolarizations. *BMC Syst. Biol.* **2017**, *11*, 42. [CrossRef] [PubMed]

28. Vo, T.; Bertam, R.; Wechselberger, M. Bifurcations of canard-induced mixed mode oscillations in a pituitary lactotroph model. *Discret. Contin. Dyn. Syst.* **2012**, *32*, 2879–2912. [CrossRef]

29. Baer, S.M.; Erneux, T.; Rinzel, J. The Slow Passage through a Hopf Bifurcation: Delay, Memory Effects, and Resonance. *SIAM J. Appl. Math.* **1989**, *49*, 55–71. [CrossRef]

30. Desroches, M.; Guckenheimer, J.; Krauskopf, B.; Kuehn, C.; Osinga, H.M.; Wechselberger, M. Mixed-Mode Oscillations with Multiple Time Scales. *SIAM Rev.* **2012**, *54*, 211–288. [CrossRef]

31. Kuehn, C. *Multiple Time Scale Dynamics*; Volume 191 of Applied Mathematical Sciences; Springer-Verlag: Heidelberg, Germany; New York, NY, USA, 2015.

32. Fenichel, N. Geometric singular perturbation theory for ordinary differential equations. *J. Differ. Equ.* **1979**, *31*, 53–98. [CrossRef]

mathematics

MDPI

Article

Existence, Uniqueness and Ulam's Stability of Solutions for a Coupled System of Fractional Differential Equations with Integral Boundary Conditions

Dimplekumar Chalishajar [1,*] and Avadhesh Kumar [2]

[1] Department of Mathematics and Computer Science, Mallory Hall, Virginia Military Institute, Lexington, VA 24450, USA
[2] School of Basic Sciences, Indian Institute of Technology Mandi, Kamand 175 005, H.P., India; soni.iitkgp@gmail.com
* Correspondence: chalishajardn@vmi.edu

Received: 2 May 2018; Accepted: 29 May 2018; Published: 7 June 2018

check for updates

Abstract: In this paper, the existence and uniqueness of the solutions to a fractional order nonlinear coupled system with integral boundary conditions is investigated. Furthermore, Ulam's type stability of the proposed coupled system is studied. Banach's fixed point theorem is used to obtain the existence and uniqueness of the solutions. Finally, an example is provided to illustrate the analytical findings.

Keywords: coupled system; green's function; integral boundary conditions; Ulam's stability

AMS Subject Classification: 34A08; 34B15; 34B27; 35B35

1. Introduction

Fractional calculus is a branch of mathematical analysis, in which arbitrary order differential and integral operators are studied. It started with a correspondence between L'Hospital and Leibnitz in 1695. Presently, plenty of literature is available on theoretical as well as numerical work on this topic. It has application in numerous fields, for example, control theory, signal and image processing, aerodynamics and biophysics [1–3]. For the fundamental concepts of fractional calculus, books like Kilbas et al. [4], Miller and Ross [5] and Halfer [6] are referred. Existence and uniqueness of solutions for fractional order differential systems in finite dimensional as well infinite dimensional spaces were studied by several authors [7–11]. Ahmad et al. [8] established existence results for nonlinear boundary value fractional integro-differential equations with integral boundary conditions.

Integral boundary conditions have several applications in real-life problems such as population dynamics, blood flow problems, underground water flow, and chemical engineering. For more details on integral boundary conditions, we refer the reader to [12]. Here we would like to consider a practical example of the integral boundary condition:

$$-x'' = g(t)f(t,x), \ \ x(0) = 0, \ \beta x'(1) = x(\eta),$$

where $t \in (0,1)$, $\eta \in (0,1]$ and β is a positive constant. This is the model for a thermostat. Solutions of the problem are stationary solutions for a one-dimensional heat equation, corresponding to a heated bar, with a controller at 1, which adds or removes heat, depending on the temperature detected by a sensor at η.

This problem can be generalized. One can consider the heat equation with nonlinear gradient source terms that vary in time. Moreover, now, the heated bar, with a controller at 1, which adds or removes heat depending on the temperature detected by sensors located at any points of the bar (it depends on how we define the function h). This problem can be written in the form

$$x'' = f(t, x, x'), \quad x(0) = 0, \quad x'(1) = \int_0^1 x(s) dh(s)(\eta).$$

Recently, Ulam's type stability has been of great interest to many researchers. In 1940, the above mentioned stability was first introduced by Ulam [13]. Then, it was explained by Hyers [14] in the subsequent years. Ulam and Hyers studied it for various kinds of differential equations with integer order. Nowadays, we describe the result of Hyers simply saying that Cauchy functional equation is Hyers-Ulam stable (or has the Hyers-Ulam stability). Next, Hyers and Ulam published some further stability results for polynomial functions, isometries, and convex functions. The Hyers' results are extended and generalized by many researchers for integer order differential equations. Plenty of significant results on Ulam's type stability can be found in the literature, we refer [15–17] and references cited therein.

To the best of our knowledge, there are only few manuscripts devoted to the study of Ulam's type stability for coupled system of fractional differential equations. Further, there is no manuscript considering the Ulam's stability for coupled system of fractional order $\alpha \in (1, 2]$ differential equations with integral boundary conditions. Motivated by this fact, in this paper, the existence, uniqueness of solutions as well as Ulam's type stability for the considered coupled system involving Caputo derivative is studied. The proposed system is given as follows:

$$
\begin{cases}
{}^c D^\alpha x(t) = f(t, y(t)), \ \alpha \in (1, 2], \ t \in \mathcal{J} \\[2mm]
{}^c D^\beta y(t) = g(t, x(t)), \ \beta \in (1, 2], \ t \in \mathcal{J} \\[2mm]
px(0) + qx'(0) = \int_0^1 a_1(x(s))ds, \quad px(1) + qx'(1) = \int_0^1 a_2(x(s))ds, \\[2mm]
\tilde{p}y(0) + \tilde{q}y'(0) = \int_0^1 \tilde{a}_1(y(s))ds, \quad \tilde{p}y(1) + \tilde{q}y'(1) = \int_0^1 \tilde{a}_2(y(s))ds,
\end{cases}
\tag{1}
$$

where $\mathcal{J} = [0, 1]$ and $f, g : \mathcal{J} \times \mathbb{R} \to \mathbb{R}$ are continuous functions. Here, $a_1, a_2, \tilde{a}_1, \tilde{a}_2 : \mathbb{R} \to \mathbb{R}$ and $p, \tilde{p} > 0$; $q, \tilde{q} \geq 0$ are real numbers.

The plan of the paper is as follows. In the second section, some useful definitions, notations, lemmas and results are given which will be required for the later sections. In the third section, existence and uniqueness of the solutions for the coupled system (1) is studied. In the fourth section, Ulam's type stability results are obtained. In the last section, a few examples are given to show the application of the obtained abstract results.

2. Preliminaries and Assumptions

In this section, some useful definitions, notations and lemmas are briefly reviewed.

Definition 1. *[4] For any function $z \in ((0, 1), \mathbb{R})$, the Caputo derivative of fractional order $\alpha \in \mathbb{R}^+$ is defined as*

$$ {}^c D^\alpha z(t) = \frac{1}{\Gamma(n - \alpha)} \int_0^t (t - s)^{n - \alpha - 1} z^{(n)}(s) ds, \quad n = [\alpha] + 1,$$

where $[\alpha]$ denotes the integer part of α and $\Gamma(\cdot)$ is the gamma function.

Definition 2. *[4] The Riemann-Liouville fractional integral of order $\alpha \in \mathbb{R}^+$ is defined by*

$$ J^\alpha z(t) = \frac{1}{\Gamma(\alpha)} \int_0^t (t - s)^{\alpha - 1} z(s) ds,$$

where $z(t) \in L^1([0,1], \mathbb{R}^+)$.

Lemma 1. *[4] For any $\alpha > 0$, then the differential equations*

$$^cD^\alpha z(t) = 0$$

has solution given by

$$z(t) = c_0 + c_1 t + c_2 t^2 + \cdots + c_{n-1} t^{n-1}, \quad c_i \in \mathbb{R}, \; i = 0, 1, \cdots, n-1,$$

where $n = [\alpha] + 1$.

Lemma 2. *[4] For any $\alpha > 0$, then the solution of the differential equations*

$$^cD^\alpha z(t) = h(t)$$

will be given by

$$J^\alpha[^cD^\alpha z(t)] = J^\alpha z(t) + c_0 + c_1 t + c_2 t^2 + \cdots + c_{n-1} t^{n-1}, \quad c_i \in \mathbb{R}, \; i = 0, 1, \cdots, n-1,$$

where $n = [\alpha] + 1$.

Lemma 3. *[8] For any $h, \gamma_1, \gamma_2 \in C([0,1], \mathbb{R})$, the unique solution of the boundary value problem*

$$\begin{cases} ^cD^\alpha z(t) = h(t), \; \alpha \in (1,2], \; t \in [0,1] \\ pz(0) + qz'(0) = \int_0^1 \gamma_1(s)ds, \quad pz(1) + qz'(1) = \int_0^1 \gamma_2(s)ds, \end{cases} \tag{2}$$

is given by

$$z(t) = \int_0^1 G_\alpha(t,s)h(s)ds + \frac{1}{p^2}\left[(p(1-t)+q)\int_0^1 \gamma_1(s)ds + (q+pt)\int_0^1 \gamma_2(s)ds\right], \tag{3}$$

where $G_\alpha(t,s)$ is the Green's function given by

$$G_\alpha(t,s) = \begin{cases} \frac{p(t-s)^{\alpha-1}+(q-pt)(1-s)^{\alpha-1}}{p\Gamma(\alpha)} + \frac{q(q-pt)(1-s)^{\alpha-2}}{p^2\Gamma(\alpha-1)}, \; s \leq t, \\ \\ \frac{(q-pt)(1-s)^{\alpha-1}}{p\Gamma(\alpha)} + \frac{q(q-pt)(1-s)^{\alpha-2}}{p^2\Gamma(\alpha-1)}, \; t \leq s. \end{cases} \tag{4}$$

Lemma 4. *The space $\mathcal{B} = \{z(t) \,|\, z \in C(\mathcal{J})\}$ is a Banach space under the defined norm $\|z\|_{\mathcal{B}} = \max_{t \in \mathcal{J}}|z(t)|$. Similarly, the norm on product space is defined by $\|(z,\tilde{z})\|_{\mathcal{B}\times\mathcal{B}} = \|z\|_{\mathcal{B}} + \|\tilde{z}\|_{\mathcal{B}}$. Obviously $(\mathcal{B} \times \mathcal{B}, \|(\cdot,\cdot)\|_{\mathcal{B}\times\mathcal{B}})$ is a Banach space. Further, the cone $\mathcal{C} \subset \mathcal{B} \times \mathcal{B}$ is defined by*

$$\mathcal{C} = \{(z,\tilde{z}) \in \mathcal{B} \times \mathcal{B} \,|\, z(t) \geq 0, \; \tilde{z}(t) \geq 0\}.$$

Here, the problem (1) is transformed into a fixed point problem. Let $\mathcal{F} : \mathcal{B} \times \mathcal{B} \to \mathcal{B} \times \mathcal{B}$ be the operator defined as

$$\mathcal{F}(x,y)(t) = \begin{pmatrix} \int_0^1 G_\alpha(t,s) f(s,y(s))ds + \frac{1}{p^2}\left[(p(1-t)+q)\int_0^1 a_1(x(s))ds \right. \\ \left. +(q+pt)\int_0^1 a_2(x(s))ds\right] \\[2mm] \int_0^1 G_\beta(t,s) g(s,x(s))ds + \frac{1}{\tilde{p}^2}\left[(\tilde{p}(1-t)+\tilde{q})\int_0^1 \tilde{a}_1(y(s))ds \right. \\ \left. +(\tilde{q}+\tilde{p}t)\int_0^1 \tilde{a}_2(y(s))ds\right] \end{pmatrix}$$

$$= \begin{pmatrix} \mathcal{F}_\alpha(y,x)(t) \\ \mathcal{F}_\beta(x,y)(t) \end{pmatrix}. \tag{5}$$

Then the fixed point of the operator $\mathcal{F}$ coincides with the solution of coupled system (1).

In order to prove the existence and uniqueness of solutions of coupled system (1), following assumptions are taken:

(A1) For $t \in \mathcal{J}$, there exist $\lambda, \mu \in C(\mathcal{J}, \mathbb{R}^+)$, such that

$$|f(t,y(t))| \leq \lambda(t) + \mu(t)|y(t)|, \ \forall \ y(t) \in C(\mathcal{J}, \mathbb{R})$$

with $\lambda^* = \sup_{t \in \mathcal{J}} \lambda(t)$, $\mu^* = \sup_{t \in \mathcal{J}} \mu(t)$.
Similarly, for $t \in \mathcal{J}$, there exist $\nu, \xi \in C(\mathcal{J}, \mathbb{R}^+)$, such that

$$|g(t,x(t))| \leq \nu(t) + \xi(t)|x(t)|, \ \forall \ x(t) \in C(\mathcal{J}, \mathbb{R})$$

with $\nu^* = \sup_{t \in \mathcal{J}} \nu(t)$, $\xi^* = \sup_{t \in \mathcal{J}} \xi(t)$.
(A2) For $t \in \mathcal{J}$, there exist positive constant L_{a_1}, L_{a_2}, such that

$$|a_1(x(t))| \leq L_{a_1}|x(t)| \text{ and } |a_2(x(t))| \leq L_{a_2}|x(t)|, \ \forall \ x(t) \in C(\mathcal{J}, \mathbb{R})$$

Similarly, For $t \in \mathcal{J}$, there exist positive constant $L_{\tilde{a}_1}$, $L_{\tilde{a}_2}$, such that

$$|\tilde{a}_1(y(t))| \leq L_{\tilde{a}_1}|y(t)| \text{ and } |\tilde{a}_2(x(t))| \leq L_{\tilde{a}_2}|y(t)|, \ \forall \ y(t) \in C(\mathcal{J}, \mathbb{R})$$

(A3)

$$\mathbb{P}_1 = \int_0^1 |G_\alpha(t,s)|\lambda(s)ds < \infty, \ \mathbb{Q}_1 = \int_0^1 |G_\alpha(t,s)|\mu(s)ds + \frac{(p+q)(L_{a_1}+L_{a_2})}{p^2} < \frac{1}{2}$$

and

$$\mathbb{P}_2 = \int_0^1 |G_\beta(t,s)|\nu(s)ds < \infty, \ \mathbb{Q}_2 = \int_0^1 |G_\beta(t,s)|\xi(s)ds + \frac{(\tilde{p}+\tilde{q})(L_{\tilde{a}_1}+L_{\tilde{a}_2})}{\tilde{p}^2} < \frac{1}{2}.$$

(A4) For all $y, \tilde{y} \in C(\mathcal{J}, R)$ and for each $t \in \mathcal{J}$ there exists a positive constant K_f, such that

$$|f(t,y) - f(t,\tilde{y})| \leq K_f|y - \tilde{y}|.$$

Similarly, for all $x, \tilde{x} \in C(\mathcal{J}, R)$ and for each $t \in \mathcal{J}$ there exists a positive constant K_g, such that

$$|g(t,x) - g(t,\tilde{x})| \leq K_g|x - \tilde{x}|.$$

(A5) For all $x, \tilde{x} \in C(\mathcal{J}, R)$ and for each $t \in \mathcal{J}$ there exist positive constants K_{a_1}, K_{a_2}, such that

$$|a_1(x) - a_1(\tilde{x})| \leq K_{a_1}|x - \tilde{x}| \text{ and } |a_2(x) - a_2(\tilde{x})| \leq K_{a_2}|x - \tilde{x}|.$$

Similarly, for all $y, \tilde{y} \in C(\mathcal{J}, R)$ and for each $t \in \mathcal{J}$ there exist positive constants $K_{\tilde{a}_1}, K_{\tilde{a}_2}$, such that

$$|\tilde{a}_1(y) - \tilde{a}_1(\tilde{y})| \le K_{\tilde{a}_1}|y - \tilde{y}| \text{ and } |\tilde{a}_2(y) - \tilde{a}_2(\tilde{y})| \le K_{\tilde{a}_2}|y - \tilde{y}|.$$

(A6) Let

(i) $\phi_1 = \mathcal{K}_{(\beta\tilde{p}\tilde{q})}K_g + K_{pq}$, where $K_{pq} = \dfrac{(p+q)(K_{a_1} + K_{a_2})}{p^2}$ and

$$
\begin{aligned}
\mathcal{K}_{(\beta\tilde{p}\tilde{q})} &= \max_{t\in[0,1]} \left| \int_0^1 G_\beta(t,s)ds \right| \\
&= \left| \int_0^t \left[\frac{\tilde{p}(t-s)^{\beta-1} + (\tilde{q}-\tilde{p}t)(1-s)^{\beta-1}}{\tilde{p}\Gamma(\beta)} + \frac{\tilde{q}(\tilde{q}-\tilde{p}t)(1-s)^{\beta-2}}{\tilde{p}^2\Gamma(\beta-1)} \right] ds \right| \\
&\quad + \left| \int_t^1 \left[\frac{(\tilde{q}-\tilde{p}t)(1-s)^{\beta-1}}{\tilde{p}\Gamma(\beta)} + \frac{\tilde{q}(\tilde{q}-\tilde{p}t)(1-s)^{\beta-2}}{\tilde{p}^2\Gamma(\beta-1)} \right] ds \right| \\
&= \frac{1}{\Gamma(\beta+1)} + \frac{2(\tilde{p}+\tilde{q})}{\Gamma(\beta+1)} + \frac{2(\tilde{q}^2+\tilde{p}\tilde{q})}{\tilde{p}^2\Gamma(\beta)};
\end{aligned}
$$

(ii) $\phi_2 = \mathcal{K}_{(\alpha pq)}K_f + K_{\tilde{p}\tilde{q}}$, where $K_{\tilde{p}\tilde{q}} = \dfrac{(\tilde{p}+\tilde{q})(K_{\tilde{a}_1} + K_{\tilde{a}_2})}{\tilde{p}^2}$ and

$$
\begin{aligned}
\mathcal{K}_{(\alpha pq)} &= \max_{t\in[0,1]} \left| \int_0^1 G_\alpha(t,s)ds \right| \\
&= \frac{1}{\Gamma(\alpha+1)} + \frac{2(p+q)}{\Gamma(\alpha+1)} + \frac{2(q^2+pq)}{p^2\Gamma(\alpha)}.
\end{aligned}
$$

3. Existence and Uniqueness Analysis

Theorem 1. *If all the assumptions (A1)–(A6) and $\phi = \max\{\phi_1, \phi_2\} < 1$ are fulfilled, then the fractional order coupled system (1) has a unique solution.*

Proof. For a positive number

$$\delta = \max\left(\frac{2\mathbb{P}_1}{1-2\mathbb{Q}_1}, \frac{2\mathbb{P}_2}{1-2\mathbb{Q}_2} \right),$$

we define a set

$$\mathcal{W} = \{(x,y) \in \mathcal{B} \times \mathcal{B} : \|(x,y)\|_{\mathcal{B}\times\mathcal{B}} \le \delta\}.$$

First, in order to prove that $\mathcal{F}$ maps $\mathcal{W}$ into itself, we have

$$
\begin{aligned}
|\mathcal{F}_\alpha(y,x)(t)| &\le \int_0^1 |G_\alpha(t,s)||f(s,y(s))|ds \\
&\quad + \frac{1}{p^2}\left[|(p(1-t)+q)| \int_0^1 |a_1(x(s))|ds + |(q+pt)| \int_0^1 |a_2(x(s))|ds \right] \\
&\le \int_0^1 |G_\alpha(t,s)|\lambda(s)ds + \int_0^1 |G_\alpha(t,s)|[\mu(s)|y(s)|]ds \\
&\quad + \frac{1}{p^2}\left[(p+q) \int_0^1 |a_1(x(s))|ds + (p+q) \int_0^1 |a_2(x(s))|ds \right] \\
&\le \int_0^1 |G_\alpha(t,s)|\lambda(s)ds + \delta\left[\int_0^1 |G_\alpha(t,s)|\mu(s)ds + \frac{(p+q)(L_{a_1}+L_{a_2})}{p^2} \right] \\
&= \mathbb{P}_1 + \delta\mathbb{Q}_1 \le \frac{\delta}{2}. \qquad (6)
\end{aligned}
$$

Now taking maximum on both side of the inequality (6) over $\mathcal{J}$, we obtain

$$\|\mathcal{F}_\alpha(y,x)\|_\mathcal{B} \leq \frac{\delta}{2}.$$

Similarly, $\|\mathcal{F}_\beta(x,y)\|_\mathcal{B} \leq \frac{\delta}{2}$. Hence, we can conclude that

$$\|\mathcal{F}(x,y)\|_{\mathcal{B}\times\mathcal{B}} \leq \delta. \tag{7}$$

Inequality (7) shows that $\mathcal{F}$ maps $\mathcal{W}$ into itself. Next, in order to show that $\mathcal{F}$ is the contraction operator when $t \in \mathcal{J}$, we have

$$
\begin{aligned}
|\mathcal{F}_\alpha(y,x)(t) - \mathcal{F}_\alpha(\bar{y},\bar{x})(t)| \quad &\leq \quad \int_0^1 |G_\alpha(t,s)||f(s,y(s)) - f(s,\bar{y}(s))|ds \\
&+ \quad \frac{1}{p^2}\Big[|(p(1-t)+q)|\int_0^1 |a_1(x(s)) - a_1(\bar{x}(s))|ds \\
&+ \quad |(q+pt)|\int_0^1 |a_2(x(s)) - a_2(\bar{x}(s))|ds\Big] \\
&\leq \quad \mathcal{K}_{(\alpha pq)}K_f|y(t) - \bar{y}(t)| \\
&+ \quad \frac{(p+q)(K_{a_1}+K_{a_2})}{p^2}|x(t) - \bar{x}(t)|.
\end{aligned}
\tag{8}
$$

When we take maximum on both side of the inequality (8) over $\mathcal{J}$, we obtain

$$\|\mathcal{F}_\alpha(y,x) - \mathcal{F}_\alpha(\bar{y},\bar{x})\|_\mathcal{B} \quad \leq \quad \mathcal{K}_{(\alpha pq)}K_f\|y - \bar{y}\|_\mathcal{B} + \frac{(p+q)(K_{a_1}+K_{a_2})}{p^2}\|x - \bar{x}\|_\mathcal{B}. \tag{9}$$

Similarly, the following can be obtained

$$\|\mathcal{F}_\beta(x,y) - \mathcal{F}_\beta(\bar{x},\bar{y})\|_\mathcal{B} \quad \leq \quad \mathcal{K}_{(\beta \tilde{p}\tilde{q})}K_g\|x - \bar{x}\|_\mathcal{B} + \frac{(\tilde{p}+\tilde{q})(K_{\tilde{a}_1}+K_{\tilde{a}_2})}{\tilde{p}^2}\|y - \bar{y}\|_\mathcal{B}. \tag{10}$$

From (9) and (9), we get

$$\|\mathcal{F}(x,y) - \mathcal{F}(\bar{x},\bar{y})\|_{\mathcal{B}\times\mathcal{B}} \quad \leq \quad \phi\|(x,y) - (\bar{x},\bar{y})\|_{\mathcal{B}\times\mathcal{B}}. \tag{11}$$

Thus, the operator $\mathcal{F}$ is strict contraction. By Banach's fixed point method, it has a unique fixed point which is the unique solution of the considered coupled system (1). □

4. Ulam's Stability Analysis

In this section, we study Ulam's type stability for the coupled system (1).

For some $\epsilon = (\epsilon_\alpha, \epsilon_\beta) > 0$, we consider the following inequality

$$
\begin{cases}
\left|{}^cD^\alpha x(t) - f(t,y(t))\right| \leq \epsilon_\alpha, & t \in \mathcal{J}, \\
\left|{}^cD^\beta y(t) - g(t,x(t))\right| \leq \epsilon_\beta, & t \in \mathcal{J}.
\end{cases}
\tag{12}
$$

The following definitions are inspired by Rus [18].

Definition 3. *The coupled system (1) is said to be Ulam-Hyers stable, if there exist $\mathcal{K}_{(\alpha\beta pq\tilde{p}\tilde{q})} = (\mathcal{K}_{(\alpha pq)}, \mathcal{K}_{(\beta\tilde{p}\tilde{q})}) > 0$ such that for every solution $(x,y) \in \mathcal{B} \times \mathcal{B}$ of the inequality (12), there exists a unique solution $(\vartheta, \kappa) \in \mathcal{B} \times \mathcal{B}$ with*

$$|(x,y)(t) - (\vartheta,\kappa)(t)| \leq \mathcal{K}_{(\alpha\beta pq\tilde{p}\tilde{q})}\epsilon, \quad t \in \mathcal{J}. \tag{13}$$

Definition 4. *The coupled system* (1) *is said to be generalized Ulam-Hyers stable, if there exist* $\Psi \in C(\mathbb{R}^+, \mathbb{R}^+)$ *with* $\Psi(0) = 0$, *such that for every solution* $(x, y) \in \mathcal{B} \times \mathcal{B}$ *of the inequality* (12), *there exist a unique solution* $(\vartheta, \kappa) \in \mathcal{B} \times \mathcal{B}$ *of the system* (1) *which satisfies*

$$|(x, y)(t) - (\vartheta, \kappa)(t)| \leq \Psi(\epsilon), \ t \in \mathcal{J}. \tag{14}$$

Remark 1. *Let* $(x, y) \in \mathcal{B} \times \mathcal{B}$ *is a solution of the system of inequality* (12) *if there exist functions* $\phi, \ \psi \in C(\mathcal{J}, \mathbb{R})$ *which depend upon x and y respectively, such that*

(R1) $|\phi(t)| \leq \epsilon_\alpha, \ |\psi(t)| \leq \epsilon_\beta, \ t \in \mathcal{J}$;
(R2) and

$$\begin{cases} {}^cD^\alpha x(t) = f(t, y(t)) + \phi(t), \ t \in \mathcal{J}, \\ {}^cD^\beta y(t) = g(t, x(t)) + \psi(t), \ t \in \mathcal{J}. \end{cases}$$

Lemma 5. *Let* $(x, y) \in \mathcal{B} \times \mathcal{B}$ *be the solution of the inequality* (12), *then the following inequality will be satisfied:*

$$\begin{cases} \left| x(t) - \int_0^1 G_\alpha(t, s) f(s, y(s)) ds - \frac{1}{p^2} \Big[(p(1-t) + q) \int_0^1 a_1(x(s)) ds \right. \\ \left. + (q + pt) \int_0^1 a_2(x(s)) ds \Big] \right| \leq \mathcal{K}_{(\alpha pq)} \epsilon_\alpha, \ t \in \mathcal{J}, \\ \left| y(t) - \int_0^1 G_\beta(t, s) g(s, x(s)) ds - \frac{1}{\tilde{p}^2} \Big[(\tilde{p}(1-t) + \tilde{q}) \int_0^1 \tilde{a}_1(y(s)) ds \right. \\ \left. + (\tilde{q} + \tilde{p}t) \int_0^1 \tilde{a}_2(y(s)) ds \Big] \right| \leq \mathcal{K}_{(\beta \tilde{p} \tilde{q})} \epsilon_\beta, \ t \in \mathcal{J}. \end{cases}$$

Proof. By Remark 1 (R2), we have

$$\begin{cases} {}^cD^\alpha x(t) = f(t, y(t)) + \phi(t), \ t \in \mathcal{J}, \\ {}^cD^\beta y(t) = g(t, x(t)) + \psi(t), \ t \in \mathcal{J}, \\ px(0) + qx'(0) = \int_0^1 a_1(x(s)) ds, \quad px(1) + qx'(1) = \int_0^1 a_2(x(s)) ds \\ \tilde{p}y(0) + \tilde{q}y'(0) = \int_0^1 \tilde{a}_1(y(s)) ds, \quad \tilde{p}y(1) + \tilde{q}y'(1) = \int_0^1 \tilde{a}_2(y(s)) ds \end{cases} \tag{15}$$

By Applying Lemma 3, the solution of (15) will be as follows:

$$\begin{cases} x(t) = \int_0^1 G_\alpha(t, s) f(s, y(s)) ds + \int_0^1 G_\alpha(t, s) \phi(s) ds + \frac{1}{p^2} \Big[(p(1-t) + q) \int_0^1 a_1(x(s)) ds \\ + (q + pt) \int_0^1 a_2(x(s)) ds \Big], \ t \in \mathcal{J}, \\ y(t) = \int_0^1 G_\beta(t, s) g(s, x(s)) ds + \int_0^1 G_\alpha(t, s) \psi(s) ds + \frac{1}{\tilde{p}^2} \Big[(\tilde{p}(1-t) + \tilde{q}) \int_0^1 \tilde{a}_1(y(s)) ds \\ + (\tilde{q} + \tilde{p}t) \int_0^1 \tilde{a}_2(y(s)) ds \Big], \ t \in \mathcal{J}. \end{cases} \tag{16}$$

Considering first equation of the system (16), we have

$$\left| x(t) - \int_0^1 G_\alpha(t,s) f(s,y(s)) ds - \frac{1}{p^2} \Big[(p(1-t)+q) \int_0^1 a_1(x(s)) ds \right.$$

$$\left. + (q+pt) \int_0^1 a_2(x(s)) ds \Big] \right| \;\leq\; \left| \int_0^1 G_\alpha(t,s) \phi(s) ds \right|$$

$$\leq\; \int_0^1 |G_\alpha(t,s)| |\phi(s)| ds.$$

By Remark 1 (R1) and using condition of (A6), we get

$$\left| x(t) - \int_0^1 G_\alpha(t,s) f(s,y(s)) ds - \frac{1}{p^2} \Big[(p(1-t)+q) \int_0^1 a_1(x(s)) ds \right.$$

$$\left. + (q+pt) \int_0^1 a_2(x(s)) ds \Big] \right| \leq \mathcal{K}_{(\alpha pq)} \epsilon_\alpha. \tag{17}$$

Repeating the same procedure for second equation of the system (16), we get

$$\left| y(t) - \int_0^1 G_\beta(t,s) g(s,x(s)) ds - \frac{1}{\tilde{p}^2} \Big[(\tilde{p}(1-t)+\tilde{q}) \int_0^1 \tilde{a}_1(y(s)) ds \right.$$

$$\left. + (\tilde{q}+\tilde{p}t) \int_0^1 \tilde{a}_2(y(s)) ds \Big] \right| \leq \mathcal{K}_{(\beta\tilde{p}\tilde{q})} \epsilon_\beta. \tag{18}$$

□

Theorem 2. *If all the assumptions (A4)–(A6) are fulfilled, then the fractional order coupled system (1) is Ulam-Hyers stable and consequently generalized Ulam-Hyers stable provided that*

$$(1 - \mathcal{K}_{pq})(1 - \mathcal{K}_{\tilde{p}\tilde{q}}) - \mathcal{K}_{\alpha pq} \mathcal{K}_{\beta\tilde{p}\tilde{q}} \neq 0.$$

Proof. Let $(x,y) \in \mathcal{B} \times \mathcal{B}$ be the solution of the system (15) and $(\vartheta, \kappa) \in \mathcal{B} \times \mathcal{B}$ be the unique solution to the following considered system:

$$\begin{cases} {}^c D^\alpha \vartheta(t) = f(t, \kappa(t)), & t \in \mathcal{J}, \\[2mm] {}^c D^\beta \kappa(t) = g(t, \vartheta(t)), & t \in \mathcal{J}, \\[2mm] p\vartheta(0) + q\vartheta'(0) = \int_0^1 a_1(\vartheta(s)) ds, & p\vartheta(1) + q\vartheta'(1) = \int_0^1 a_2(\vartheta(s)) ds \\[2mm] \tilde{p}\kappa(0) + \tilde{q}\kappa'(0) = \int_0^1 \tilde{a}_1(\kappa(s)) ds, & \tilde{p}\kappa(1) + \tilde{q}\kappa'(1) = \int_0^1 \tilde{a}_2(\kappa(s)) ds \end{cases} \tag{19}$$

Using Lemma 3, the solution of (19)

$$\begin{cases} \vartheta(t) = \int_0^1 G_\alpha(t,s) f(s, \kappa(s)) ds + \frac{1}{p^2} \Big[(p(1-t)+q) \int_0^1 a_1(\vartheta(s)) ds \\[2mm] \qquad + (q+pt) \int_0^1 a_2(\vartheta(s)) ds \Big], \quad t \in \mathcal{J}, \\[4mm] \kappa(t) = \int_0^1 G_\beta(t,s) g(s, \vartheta(s)) ds + \frac{1}{\tilde{p}^2} \Big[(\tilde{p}(1-t)+\tilde{q}) \int_0^1 \tilde{a}_1(\kappa(s)) ds \\[2mm] \qquad + (\tilde{q}+\tilde{p}t) \int_0^1 \tilde{a}_2(\kappa(s)) ds \Big], \quad t \in \mathcal{J}. \end{cases} \tag{20}$$

We have

$$
\begin{aligned}
|x(t) - \vartheta(t)| \quad = \quad & \left| x(t) - \int_0^1 G_\alpha(t,s) f(s, \kappa(s)) ds - \frac{1}{p^2} \Big[(p(1-t) + q) \int_0^1 a_1(\vartheta(s)) ds \right. \\
& + \left. (q + pt) \int_0^1 a_2(\vartheta(s)) ds \Big] \right| \\
\leq \quad & \left| x(t) - \int_0^1 G_\alpha(t,s) f(s, y(s)) ds - \frac{1}{p^2} \Big[(p(1-t) + q) \int_0^1 a_1(x(s)) ds \right. \\
& + \left. (q + pt) \int_0^1 a_2(x(s)) ds \Big] \right| \\
& + \left| \int_0^1 G_\alpha(t,s) f(s, y(s)) ds - \int_0^1 G_\alpha(t,s) f(s, \kappa(s)) ds \right| \\
& + \left| \frac{1}{p^2} \Big[(p(1-t) + q) \int_0^1 a_1(x(s)) ds + (q + pt) \int_0^1 a_2(x(s)) ds \Big] \right. \\
& - \left. \frac{1}{p^2} \Big[(p(1-t) + q) \int_0^1 a_1(\vartheta(s)) ds + (q + pt) \int_0^1 a_2(\vartheta(s)) ds \Big] \right| \\
\leq \quad & \mathcal{K}_{\alpha pq} \epsilon_\alpha + \mathcal{K}_{\alpha pq} |y(t) - \kappa(t)| + K_{pq} |x(t) - \vartheta(t)|,
\end{aligned}
$$

where $\mathcal{K}_{\alpha pq} = \mathcal{K}_{(\alpha pq)} K_f$.

Hence, we get

$$
(1 - K_{pq}) \|x - \vartheta\|_{\mathcal{B}} \leq \mathcal{K}_{(\alpha pq)} \epsilon_\alpha + \mathcal{K}_{\alpha pq} \|y - \kappa\|_{\mathcal{B}}. \tag{21}
$$

Similarly, we have

$$
(1 - K_{\bar{p}\bar{q}}) \|y - \kappa\|_{\mathcal{B}} \leq \mathcal{K}_{(\beta \bar{p}\bar{q})} \epsilon_\beta + K_{\beta \bar{p}\bar{q}} \|x - \vartheta\|_{\mathcal{B}}, \tag{22}
$$

where $K_{\beta \bar{p}\bar{q}} = \mathcal{K}_{(\beta \bar{p}\bar{q})} K_g$.

From (21) and (22), it can be written as

$$
\begin{cases}
(1 - K_{pq}) \|x - \vartheta\|_{\mathcal{B}} - \mathcal{K}_{\alpha pq} \|y - \kappa\|_{\mathcal{B}} \leq \mathcal{K}_{(\alpha pq)} \epsilon_\alpha \\
(1 - K_{\bar{p}\bar{q}}) \|y - \kappa\|_{\mathcal{B}} - K_{\beta \bar{p}\bar{q}} \|x - \vartheta\|_{\mathcal{B}} \leq \mathcal{K}_{(\beta \bar{p}\bar{q})} \epsilon_\beta
\end{cases} \tag{23}
$$

The matrix representation of (23) is as follows

$$
\begin{pmatrix} (1 - K_{pq}) & -\mathcal{K}_{\alpha pq} \\ -K_{\beta \bar{p}\bar{q}} & (1 - K_{\bar{p}\bar{q}}) \end{pmatrix} \begin{pmatrix} \|x - \vartheta\|_{\mathcal{B}} \\ \|y - \kappa\|_{\mathcal{B}} \end{pmatrix} \leq \begin{pmatrix} \mathcal{K}_{(\alpha pq)} \epsilon_\alpha \\ \mathcal{K}_{(\beta \bar{p}\bar{q})} \epsilon_\beta \end{pmatrix}.
$$

After simplification of the above inequality, we have

$$
\begin{pmatrix} \|x - \vartheta\|_{\mathcal{B}} \\ \|y - \kappa\|_{\mathcal{B}} \end{pmatrix} \leq \begin{pmatrix} \frac{(1 - K_{\bar{p}\bar{q}})}{\Delta} & \frac{K_{\alpha pq}}{\Delta} \\ \frac{K_{\beta \bar{p}\bar{q}}}{\Delta} & \frac{(1 - K_{pq})}{\Delta} \end{pmatrix} \begin{pmatrix} \mathcal{K}_{(\alpha pq)} \epsilon_\alpha \\ \mathcal{K}_{(\beta \bar{p}\bar{q})} \epsilon_\beta \end{pmatrix},
$$

where $\Delta = (1 - K_{pq})(1 - K_{\bar{p}\bar{q}}) - K_{\alpha pq} K_{\beta \bar{p}\bar{q}} \neq 0$.

Further simplification gives

$$
\|x - \vartheta\|_{\mathcal{B}} \leq \frac{(1 - K_{\bar{p}\bar{q}}) \mathcal{K}_{(\alpha pq)} \epsilon_\alpha}{\Delta} + \frac{K_{\alpha pq} \mathcal{K}_{(\beta \bar{p}\bar{q})} \epsilon_\beta}{\Delta} \tag{24}
$$

$$
\|y - \kappa\|_{\mathcal{B}} \leq \frac{K_{\beta \bar{p}\bar{q}} \mathcal{K}_{(\alpha pq)} \epsilon_\alpha}{\Delta} + \frac{(1 - K_{pq}) \mathcal{K}_{(\beta \bar{p}\bar{q})} \epsilon_\beta}{\Delta} \tag{25}
$$

From inequalities (24) and (25), we have

$$
\begin{aligned}
\|x - \vartheta\|_{\mathcal{B}} + \|y - \kappa\|_{\mathcal{B}} \quad &\leq \quad \frac{(1 - K_{\tilde{p}\tilde{q}})\mathcal{K}_{(\alpha p q)}\epsilon_{\alpha}}{\Delta} + \frac{K_{\alpha p q}\mathcal{K}_{(\beta\tilde{p}\tilde{q})}\epsilon_{\beta}}{\Delta} + \frac{K_{\beta\tilde{p}\tilde{q}}\mathcal{K}_{(\alpha p q)}\epsilon_{\alpha}}{\Delta} \\
&\quad + \frac{(1 - K_{pq})\mathcal{K}_{(\beta\tilde{p}\tilde{q})}\epsilon_{\beta}}{\Delta}
\end{aligned}
$$

Therefore, we have

$$
\|(x, y) - (\vartheta, \kappa)\|_{\mathcal{B} \times \mathcal{B}} \leq \mathcal{K}_{(\alpha\beta p q \tilde{p}\tilde{q})}\epsilon, \tag{26}
$$

where $\epsilon = \max\{\epsilon_{\alpha}, \epsilon_{\beta}\}$ and

$$
\mathcal{K}_{(\alpha\beta p q \tilde{p}\tilde{q})} = \frac{(1 - K_{\tilde{p}\tilde{q}})\mathcal{K}_{(\alpha p q)}}{\Delta} + \frac{K_{\alpha p q}\mathcal{K}_{(\beta\tilde{p}\tilde{q})}}{\Delta} + \frac{K_{\beta\tilde{p}\tilde{q}}\mathcal{K}_{(\alpha p q)}}{\Delta} + \frac{(1 - K_{pq})\mathcal{K}_{(\beta\tilde{p}\tilde{q})}}{\Delta}.
$$

Hence, by inequality (26), we can conclude that the coupled system (1) is Ulam-Hyers stable. Further, inequality (26) can be written as

$$
\|(x, y) - (\vartheta, \kappa)\|_{\mathcal{B} \times \mathcal{B}} \leq \Psi(\epsilon), \text{ where } \Psi(0) = 0. \tag{27}
$$

By inequality (27), we further conclude that the coupled system (1) is generalized Ulam-Hyers stable. □

5. Application

Example 1. We consider the following fractional order coupled system:

$$
\begin{cases}
{}^{c}D^{\alpha}x(t) = \frac{1}{(t+7)^2}\frac{|y(t)|}{1+|y(t)|}, \ \alpha \in (1, 2], \ t \in \mathcal{J} = [0, 1] \\[2mm]
{}^{c}D^{\beta}y(t) = \frac{1}{100}\big[t\cos x(t) + x(t)\sin t\big], \ \beta \in (1, 2], \ t \in \mathcal{J} = [0, 1] \\[2mm]
x(0) + x'(0) = \int_0^1 \frac{|x(s)|}{13+|x(s)|}ds, \quad x(1) + x'(1) = \int_0^1 \frac{|x(s)|}{15+|x(s)|}ds \\[2mm]
y(0) + y'(0) = \int_0^1 \frac{1}{26}\big[\cos y(s) + \sin y(s)\big]ds, \quad y(1) + y'(1) = \int_0^1 \frac{1}{30}\big[\cos y(s) + \sin y(s)\big]ds
\end{cases} \tag{28}
$$

By comparing the coupled systems (28) to (1), the following values are derived:

$$
p = q = \tilde{p} = \tilde{q} = 1, \ K_{a_1} = K_{\tilde{a}_1} = \frac{1}{13} \text{ and } K_{a_2} = K_{\tilde{a}_2} = \frac{1}{15}.
$$

Here,

$$
f(t, y(t)) = \frac{1}{(t+7)^2}\frac{|y(t)|}{1+|y(t)|} \text{ and } g(t, x(t)) = \frac{1}{100}\big[t\cos x(t) + x(t)\sin t\big].
$$

As, $|f(t, y) - f(t, \tilde{y})| \leq \frac{1}{49}|y - \tilde{y}|$ and $|g(t, x) - g(t, \tilde{x})| \leq \frac{1}{50}|x - \tilde{x}|$, therefore (A4) is satisfied with $K_f = \frac{1}{49}$ and $K_g = \frac{1}{50}$. Further, we have

$$
\begin{aligned}
\phi_1 &= \left(\frac{1}{\Gamma(\beta+1)} + \frac{2(\tilde{p}+\tilde{q})}{\Gamma(\beta+1)} + \frac{2(\tilde{q}^2 + \tilde{p}\tilde{q})}{\tilde{p}^2\Gamma(\beta)}\right)K_g + \frac{(p+q)(K_{a_1} + K_{a_2})}{p^2} \\[2mm]
&= \left(\frac{1}{\Gamma(\beta+1)} + \frac{4}{\Gamma(\beta+1)} + \frac{4}{\Gamma(\beta)}\right)\frac{1}{50} + 2\left(\frac{1}{13} + \frac{1}{15}\right) \\[2mm]
&< 1.
\end{aligned}
$$

Similarly, $\phi_2 < 1$. Hence, the coupled system (28) has a unique solution. Moreover, the condition $(1 - K_{pq})(1 - K_{\bar{p}\bar{q}}) - K_{\alpha pq}K_{\beta\bar{p}\bar{q}} \neq 0$ in Theorem 2 is also satisfied. Therefore, coupled system (28) is Ulam-Hyers stable as well as generalized Ulam-Hyers stable.

6. Conclusions

Here we have studied the existence and uniqueness of the solutions as well as the stability for a coupled system of fractional order $\alpha \in (1, 2]$ differential equation with integral boundary conditions. We have discussed two types of stability, called Ulam-Hyers stability and generalized Ulam-Hyers stability. As a future work, one can generalize the same concept of stability to a neutral time delay system/inclusion as well as state delay system/inclusion (finite and infinite delay), which have some useful scientific applications. This will enhance a new direction of research: a special kind of phase space to be used for the study of controllability and stability of an infinite delay system/inclusion.

Author Contributions: Both the authors have contributed equally in the article.

Acknowledgments: We authors are thankful to referees for their valuable suggestions and comments which greatly improved the manuscript.

Conflicts of Interest: The authors declare no conflicts of interest.

References

1. Muslim, M.; Kumar, A. Controllability of fractional differential equation of order $\alpha \in (1, 2]$ with non-instantaneous impulses. *Asian J. Control* **2017**. [CrossRef]
2. Yang, Q.; Chen, D. Fractional calculus in image processing: A review. *Fract. Calcul. Appl. Anal.* **2016**, *19*, 1222–1249. [CrossRef]
3. Magin, R.L. *Fractional Calculus in Bioengineering*; Begell House: Redding, CA, USA, 2006.
4. Kilbas, A.A.; Srivastava, H.M.; Trujillo, J.J. *Theory and Applications of Fractional Differential Equations*; North-Holland Mathematics Studies; Elsevier Science Inc.: New York, NY, USA, 2006; Volume 204.
5. Miller, K.S.; Ross, B. *An Introduction to Fractional Calculus and Fractional Differential Equations*; A Wiley Interscience Publication, Wiley: New York, NY, USA, 1993.
6. Hilfer, R. *Applications of Fractional Calculus in Physics*; World Scientific: Singapore, 2000.
7. Khan, A.; Shah, K.; Li, Y.; Khan, T.S. Ulam type stability for a coupled system of boundary value problems of nonlinear fractional differential. *Equ. J. Funct. Spaces* **2017**, *2*, 1–8. [CrossRef]
8. Ahmad, B.; Nieto, J.J. Existence results for nonlinear boundary value problems of fractional integro-differential equations with integral boundary conditions. *Bound. Value Probl.* **2009**, *2009*. [CrossRef]
9. Muslim, M.; Kumar, A.; Agarwal, R.P. Exact controllability of fractional integro-differential systems of order $\alpha \in (1, 2]$ with deviated argument. *Analele Universităţii Oradea* **2017**, *XXIV*, 185–194.
10. Chalishajar, D.N.; Karthikeyan, K. Boundary value Problems for Impulsive Fractional Evolution Integrodifferential Equations with Gronwall's inequality in Banach spaces. *J. Discontinuity Nonlinearity Complex.* **2014**, *3*, 33–48. [CrossRef]
11. Chalishajar, D.N.; Karthikeyan, K. Existence and Uniqueness Results for Boundary Value Problems of Higher order Fractional Integro-differential Equations Involving Gronwall's Inequality in Banach spaces. *Acta Math. Sci. Ser. A* **2013**, *33B*, 758–772. [CrossRef]
12. Ahmad, B.; Alsaedi, A.; Alghamdi, B.S. Analytic approximation of solutions of the forced Duffing equation with integral boundary conditions. *Nonlinear Anal. Real World Appl.* **2008**, *9*, 1727–1740. [CrossRef]
13. Ulam, S.M. *A Collection of the Mathematical Problems*; Interscience: New York, NY, USA, 1960.
14. Hyers, D.H. On the stability of the linear functional equation. *Proc. Natl. Acad. Sci. USA* **1941**, *27*, 222–224. [CrossRef] [PubMed]
15. Jung, S.M. Hyers-Ulam stability of linear differential equations of first order. *Appl. Math. Lett.* **2006**, *19*, 854–858. [CrossRef]
16. Wang, J.; Feckan, M. A general class of impulsive evolution equations. *Topol. Methods Nonlinear Anal.* **2015**, *46*, 915–933. [CrossRef]

17. Muslim, M.; Kumar, A.; Fečkan, M. Existence, uniqueness and stability of solutions to second order nonlinear differential equations with non-instantaneous impulses. *J. King Saud Univ. Sci.* **2016**. [CrossRef]
18. Rus, I.A. Ulam stabilities of ordinary differential equations in a Banach space. *Carpath. J. Math.* **2010**, *26*, 103–107.

mathematics

MDPI

Correction

Correction: Kuniya, T. Stability Analysis of an Age-Structured SIR Epidemic Model with a Reduction Method to ODEs. *Mathematics* 2018, 6, 147

Toshikazu Kuniya[ID]

Graduate School of System Informatics, Kobe University, 1-1 Rokkodai-cho, Nada-ku, Kobe 657-8501, Japan; tkuniya@port.kobe-u.ac.jp

Received: 13 November 2018; Accepted: 14 November 2018; Published: 15 November 2018

check for updates

I have found an error in Equation (17) in my paper [1], and thus I would like to make the following correction: On page 7, Equation (17) should be changed from $\Delta = a_1 a_2 a_3 - a_0 a_3^2 - a_1^3 > 0$ to the following correct version: $\Delta = a_1 a_2 a_3 - a_0 a_3^2 - a_1^2 > 0$. I apologize for any inconvenience caused to the readers. This change does not affect the scientific results.

References

1. Kuniya, T. Stability Analysis of an Age-Structured SIR Epidemic Model with a Reduction Method to ODEs. *Mathematis* **2018**, *6*, 147. [CrossRef]

MDPI

St. Alban-Anlage 66

4052 Basel

Switzerland

Tel. +41 61 683 77 34

Fax +41 61 302 89 18

www.mdpi.com

Mathematics Editorial Office

E-mail: mathematics@mdpi.com

www.mdpi.com/journal/mathematics